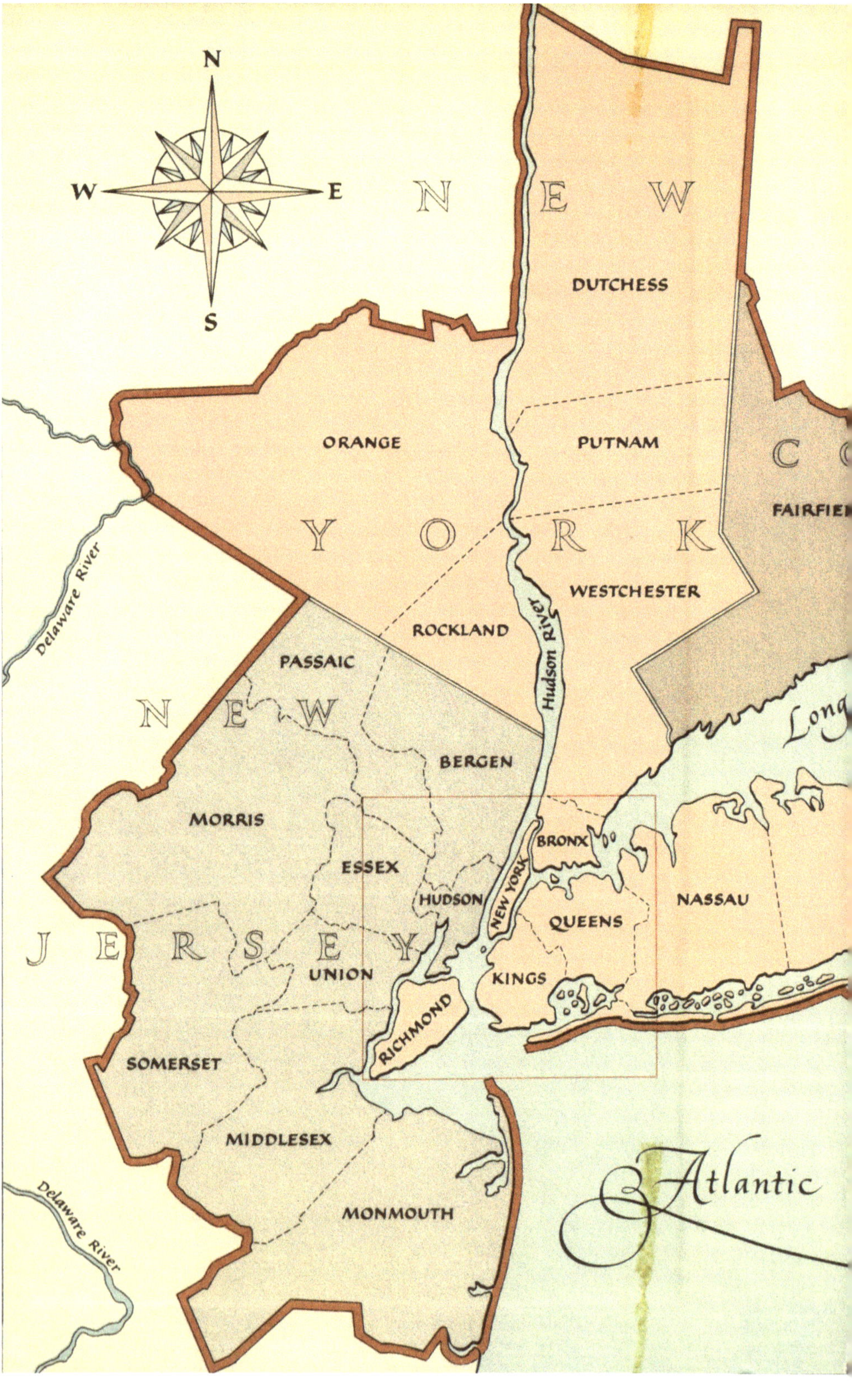

N
W
E
S
NEW YORK
DUTCHESS
ORANGE
PUTNAM
C
FAIRFIE
WESTCHESTER
ROCKLAND
Delaware River
PASSAIC
NEW JERSEY
Hudson River
Long
BERGEN
MORRIS
BRONX
ESSEX
NEW YORK
HUDSON
NASSAU
QUEENS
UNION
KINGS
RICHMOND
SOMERSET
MIDDLESEX
Atlantic
MONMOUTH
Delaware River

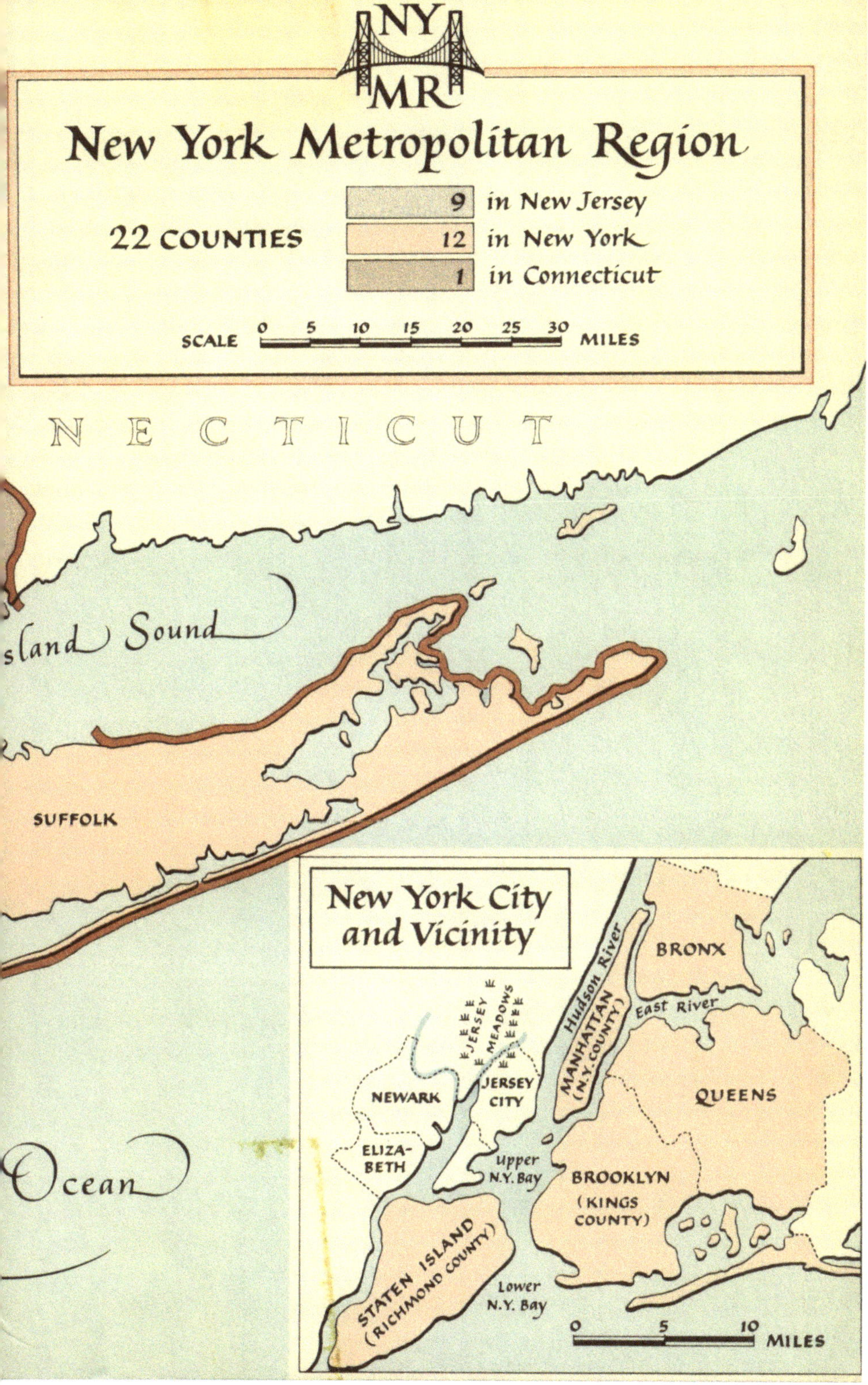
NY
MR
New York Metropolitan Region
22 COUNTIES
9 in New Jersey
12 in New York
1 in Connecticut
SCALE 0 5 10 15 20 25 30 MILES
NECTICUT
sland Sound
SUFFOLK
Ocean
New York City and Vicinity
BRONX
Hudson River
MANHATTAN (N.Y. COUNTY)
East River
JERSEY MEADOWS
NEWARK
JERSEY CITY
QUEENS
ELIZA-BETH
Upper N.Y. Bay
BROOKLYN (KINGS COUNTY)
STATEN ISLAND (RICHMOND COUNTY)
Lower N.Y. Bay
0 5 10 MILES

NEW YORK METROPOLITAN REGION STUDY

RAYMOND VERNON, DIRECTOR

A STUDY UNDERTAKEN BY THE GRADUATE SCHOOL OF PUBLIC ADMINISTRATION, HARVARD UNIVERSITY, FOR REGIONAL PLAN ASSOCIATION, INC.

Max Hall, Editorial Director

MADE IN NEW YORK

CASE STUDIES IN METROPOLITAN MANUFACTURING

Max Hall, editor

Studies by
Roy B. Helfgott
W. Eric Gustafson
James M. Hund

HARVARD UNIVERSITY PRESS
Cambridge, Massachusetts · 1959

Endpaper map by Jeanyee Wong

Charts by H. I. Forman

Designed by Marcia R. Lambrecht

Library of Congress Catalog Card Number 59-10316
Printed in the United States of America

Foreword

This is one of a series of books on the forces that shape metropolitan areas. In particular, the series has to do with the forces that shape the largest and most complex metropolitan area in the United States, a 22-county expanse which takes in parts of three states but which, for convenience, we have termed the New York Metropolitan Region.

In 1956, the Regional Plan Association, Inc., a nonprofit research and planning agency whose purpose is to promote the coordinated development of these 22 counties, requested the Graduate School of Public Administration of Harvard University to undertake a three-year study of the Region. The challenging task was to analyze the key economic and demographic features of the Region and to project them to 1965, 1975, and 1985.

The resulting studies are reports to the Regional Plan Association. At the same time, they are designed to be of service to a much broader audience. Most Americans now live in metropolitan areas; indeed ever-increasing proportions of the world's populations are gravitating to metropolitan clusters. Their well-being depends to a considerable extent on how these areas develop. Yet the scholar's understanding of the currents underlying the rise of such areas seems grossly inadequate.

As a study of these underlying currents, this project is neither a blueprint for action nor an analysis of metropolitan government. It has no recommendations to make about the physical structure of the Region or about the form or activities of the governmental bodies there. At the same time, it is a necessary prelude to future planning studies of the Region and to well considered recommendations for governmental action. Its end product is an analysis of the Region's probable development, assuming that the economic and demographic forces in sight follow their indicated course and assuming that the role of government is largely limited to existing policies.

The results of the Study, it is hoped, will be applied in many ways. Governments and enterprises in the Region should be in a better position to plan their future programs if they become more closely aware of the economic environment in which they may expect to operate. Other metropolitan areas, it is already evident, will benefit from the methodology and the conclusions which the Study has developed.

From the first, there has been a general recognition that the main part of the Study would have to be done by a group located within the New York Metropolitan Region and devoted exclusively to the project. Such a group was assembled in New York. The work that followed was a complex partnership. The New York staff functioned in close harness with members of the Harvard University faculty. It drew on the faculties of other universities, including Columbia University, Fordham University, Hofstra College, New York University, and Rutgers University. It obtained the help of dozens of governmental organizations in the Region, and literally hundreds of private groups and individuals. It made use of the materials which the Regional Plan Association had painstakingly pulled together in prior years.

Each book in the series has a place in the total structure of the Study; yet each is designed to be a complete work in itself. The final report, containing the synthesis and projections of the Study, is scheduled for publication sometime in 1960.

It is not easy to account for all the elements that went into the making of this book nor of the others in the series. The Regional Plan Association performed an indispensable function in conceiving and sponsoring the idea of a study. The Ford Foundation and the Rockefeller Brothers Fund generously provided the financial support. The usual formula in such a situation obviously applies: credit for the Study's results must be shared with those who helped to bring it about, but the onus of error or omission lies with us.

The several volumes in the series bear the names of their principal authors. The undertaking as a whole has been under the direction of Raymond Vernon. He is responsible for the final report and substantial parts of other studies, and his guidance is evident throughout the series.

EDWARD S. MASON
for The Graduate School
of Public Administration,
Harvard University

Contents

Tables

WOMEN'S AND CHILDREN'S APPAREL

PRINTING AND PUBLISHING

ELECTRONICS

APPENDIX B

APPENDIX D

APPENDIX E

Charts

WOMEN'S AND CHILDREN'S APPAREL

ELECTRONICS

Introduction
THREE INDUSTRIES ON THE MOVE

By
Max Hall

Three Industries on the Move

The New York Metropolitan Region, spreading over the face of New Jersey and New York State and taking a bite out of Connecticut,* has nearly two million people working in manufacturing establishments. They make up about 28 per cent of the Region's total employment. In New York City itself, which is hardly the popular image of an industrial center, there are nearly a million manufacturing employees. More than half a million of these work in Manhattan south of Central Park. Broadway is a great manufacturing street, though it may not look it.

By magnitude alone, therefore, manufacturing is bound to play a crucial role in the Region's future development. But magnitude is not the only reason why this is so. There is also the fact that manufacturing has roving habits. Certain other economic activities, including retail trade and public utilities, stick close to their local markets. But manufacturing is always shifting on the map.

It shifts whenever a firm migrates, whenever a new plant starts production, whenever an old one closes down. It also shifts by means of faster growth in some places than others. In each industry all the migrations, births, deaths, and differing rates of growth add up to net changes in its locational pattern. As the industry thus travels, each geographical area experiences net changes both in its

* Of the Region's 22 counties, New York City accounts for five: New York County (Manhattan), Bronx, Queens, Kings (Brooklyn), and Richmond. Seven others are also in New York State: Westchester, Rockland, Orange, Dutchess, Putnam, Nassau, and Suffolk. Nine of the counties are in New Jersey: Hudson, Union, Essex, Passaic, Bergen, Monmouth, Middlesex, Somerset, and Morris. One county, Fairfield, is in Connecticut.

number of employees and in its share of the industry's national employment. If the industry is expanding nationally, as industries generally are, the area's share may decline even while the area's absolute employment in the industry continues to increase. This is a familiar happening in the industrial history of the New York Metropolitan Region.

Some manufacturing, it is true, is almost as tightly tied to local populations as retail trade; witness the baking of bread, the bottling of soft drinks, and the production of daily newspapers. But most manufacturing is performed for the national market and is under no compulsion to locate next door to a limited group of consumers. In the New York Metropolitan Region, for example, about seven-eighths of the manufacturing employees are engaged in making the kinds of goods that are distributed nationally. Some of those goods of course are consumed in the Region itself (the population is about 16 million), but about two-thirds of the Region's manufacturing employees are producing for destinations outside the Region entirely.

The forces that cause national-market manufacturing to shift are as complex as they are important. One approach to an understanding of those forces is to analyze the manufacturing economy of the Region as a whole, look for groups of industries that are shifting location, and try to find some meaningful generalizations about the rising and declining groups. Another approach is to trace the impact of significant locational factors such as labor or transport or space or taxes.

Those approaches are important, and they will be used in other volumes of the New York Metropolitan Region Study. But by themselves they would not be enough. In order to see how a given slice of manufacturing responds to many interacting forces, and in order to make sure that we do not miss some of the more subtle forces altogether, we need to select particular manufacturing industries and put them under a microscope. That is the approach taken in this volume.

❡ The industries we chose

Our three case studies are on the following manufacturing industries:

Women's and children's apparel. We omitted things worn on the head and feet, and a few articles in between, but even so, the industry employs about 260,000 people in the Region.

Printing and publishing. Many a newspaper reporter or other publishing employee is unaware of the fact that in the eyes of the Census Bureau he is a manufacturing worker. But newspapers, magazines, books, and other printed materials are unquestionably manufactured, and those who prepare them are considered as having a hand in the process. The industry employs more than 160,000 in the Region.

Electronics. This is the elusive one. It grows fast and creates new products often. By the latest estimate, a rough one at best, about 98,000 were working in the Region's electronics plants in 1956. The number is surely much larger today.

We chose these industries because of their importance to the Region and the Region's importance to *them.* The three employment figures add to a total of 518,000, which is about 28 per cent of the manufacturing employees of the Region. The Region has been overwhelmingly the dominant center of the nation for women's and children's apparel and for printing and publishing since well before the beginning of the twentieth century. And it is one of the chief centers of electronics.

Variety is a characteristic of the Region's manufacturing economy in general, and the selected industries provide plenty of variety. Indeed, though we speak of three industries, we can just as well regard them as twelve, or twenty, or more. Thus it is best to think of each case study as having to do with a group of related industries. Coats and suits present a different set of problems from pajamas or pedal-pushers; maids' uniforms from party dresses; brassieres from blouses. There are many kinds of books, and all books differ from newspapers; producing *Life* magazine is not like producing looseleaf notebooks; a piece of sheet music scarcely resembles a get-well-quick greeting card. Transistor radios, automatic controls for landing airplanes in foggy weather, and guidance systems for moon missiles are only a few of the heterogeneous array of things we call electronic products. Moreover, even if one considers a single product, like a dress, or a magazine, or a navigational system for a submarine, the designing differs from the sewing, the publishing from the printing, the making of components from the final assembly.

The industry groups differ strikingly in the characteristics of demand for their products; the total demand ranges from a high degree of predictability, as in the case of newspapers, to a hopeless inability to predict, as in the case of military electronics. They differ in the emphasis on technological advance, with apparel at the low end of the scale and electronics at the high end. They have a wide variety of labor needs and transport needs, and they range in their space needs from cubby-holes in loft buildings to low-slung plants longer than a football field.

Nevertheless, despite the impressive diversity between and within the industry groups, the case studies make it clear that the New York Metropolitan Region has had similar experiences with all three.

In each industry group the manufacture of some products and the performance of certain stages of manufacture have tended to shift markedly away from the Region. These slices of manufacturing have found other places more conductive to growth. This does not mean they have abandoned the Region entirely, but they have expanded faster in other places, say the Midwest, or the anthracite regions of Pennsylvania, or the South, or New England, or Southern California. On the other hand, in every industry group some products and some stages of manufacture have found the economic climate of New York and vicinity so healthful that they have tended to shift into the Region, or, if the Region has lost ground relative to the rest of the country, it has lost ground only slowly. Finally—this is the critical point—the activities that have tended to shift outward from the Region have had much in common; and the activities showing the strongest propensity for the Region have also had common traits.

A GOOD PLACE TO VISIT, BUT . . .

What were the outward-bound activities? At the risk of oversimplifying the elaborate analyses given by the authors of our case studies, we can summarize them as follows.

The case study on apparel shows that it was the sewing, as distinguished from the design and merchandising, that tended to flee the Region. At first it was the sewing of the simpler garments, such as household dresses, undergarments, and children's clothing. These were the products which were less affected by style changes and whose specifications could be more accurately anticipated. In other words, they were more standardized. They were the products that lent themselves most readily to an elaborate division of labor, with each worker having only one portion of the garment to sew. Relatively little skill was involved. The numbers of identical garments produced were relatively great. In recent years, the Region has also been losing in the production of more style-oriented garments such as dresses priced by the unit (rather than by the dozen), but even in these it was the cheaper lines that tended to drift away.

In printing and publishing, the Region's losses were again in the production end. That is, they were in the press-work, binding, and mailing of certain products. The publishing activities—the gathering, writing, drawing, or editing of the material to be printed—showed no inclination to leave. The bulk of what departed was standardized work, standardized in its press-work stage, though of course not in its content. In general it was the kind of work that is done in large batches, or long "runs," even into the millions of copies.

Finally, the study of the electronics group shows that the Region's weaknesses were again in the most standardized products. The Region did not retain a very strong grip on electronic items produced in large volume for a mass market. These were notably radio and television sets, ordinary vacuum tubes and other standardized components. The exodus of the standardized products commenced early in the history of the industry, at the time when radio technology began to settle down and what had been an uncertain and chaotic industry was leaving the small shop and taking to the assembly line. As the electronics industry expanded, many firms producing standardized items never saw New York at all, but came into being in other areas.

The chief common denominator in these manufacturing operations that were attracted more strongly to other places than to the Region appears to be standardization. The rest of the country gained relative to New York in products whose specifications could be planned in advance with reasonable assurance. Large numbers of identical copies—house dresses, magazines, radio sets—could be poured out of the plants without making any changes in the design.

These generalizations do not mean, of course, that all the standardized operations in the three industry groups bypassed the Region or migrated from it. The printing of newspapers, as distinguished from the publishing, is a standardized operation; yet New York's papers continue to be printed at the heart of the city, just as other American papers do in *their* cities. Their location illustrates a phenomenon that we have already mentioned, the necessity for local-market industries to stick close to their markets. Besides, newspaper printing has to be done so fast that, unlike the printing of magazines and books, it has to be performed close to the scene of the publishing. But the fact remains that the manufacture of standardized products for *national markets* has shown pervasive tendencies, in our three industry groups, to prefer locations far from New York.

What did those locations have that the New York Metropolitan Region did not?

One thing was cheaper labor—especially unskilled labor. Most places in the nation, and especially small cities, semi-rural areas, and areas of surplus labor supply, could provide workers at lower cost than the New York Metropolitan Region. The wage rates were generally lower; besides, the labor costs imposed by restrictive working rules, and by employee benefits other than wages, were likely to be less. So were the losses due to labor-management troubles. The lower cost of labor played a powerful part in the shifting of the more standardized portions of the apparel industry to Pennsylvania, Massachusetts, and the South. It did the same in the shifting of the more repetitive kinds of electronics manufacturing to New England, the South, and small cities of the Midwest. It was also instrumental in the spread of book printing up and down the eastern part of the country, and in the movement of some of New York City's job printing to places within overnight trucking distance. It had a part, though not a controlling one, in the migration of magazine printing to the Midwest.

The second outstanding advantage that other areas had over the New York Metropolitan Region was the lower cost of transportation. Here we are not talking about the cost of obtaining materials, for in all three of our industry groups this cost has not been sufficiently variable from region to region to prompt much of the shifting that has gone on. We mean the cost of shipping the final product. There has never been a decade in the history of the United States when the population center of the country did not move westward. For a long time now the central points for both population and income have been in the Midwest, and it is cheaper and faster to serve the national market from the general vicinity of Chicago than from anywhere else. This advantage was one of the reasons why radio and television manufacturing shifted heavily to the Midwest. It was the principal force that drew the printing of magazines to the middle of the nation. On the whole, it did not affect the location of manufacturing so pervasively as did the labor-cost advantage, but its influence was considerable.

Those two cost factors, labor and transport, are repeatedly emphasized in our case studies. Other factors have also injected themselves—for example, the cost of space and the cost of taxes. In the apparel industries, space has been more important than transportation.

Nevertheless, such cost factors as these, strong as they are, do not furnish us with a full explanation of the shifts of standardized manufacturing away from the Region. The case studies show that in the industries under examination the attraction exerted by opportunities for cost reductions became effective only when and if some kind of manufacturing activity became ripe for the move. And the ripening process usually consisted of changes in the activity itself, or changes in the circumstances in which it was carried on. But before discussing those changes, we have to describe the currents that ran counter to the outward movement.

THE LURE OF THE METROPOLIS

We said earlier that while some national-market manufacturing activities were shifting markedly away from the New York Metropolitan Region, others were drawn to the Region or at least showed little tendency to be drawn away.

In the women's and children's apparel industries, these activities with the strongest propensity for the Region tended to be of the following kinds: the designing of garments; the merchandising of garments; certain highly-skilled operations like the cutting of the fabric; certain sewing operations, especially those involving the more expensive, more style-oriented garments, and those performed under the tailor system, in which one skilled tailor sews almost the entire garment; and finally the operations of the smaller shops of all descriptions, the shops that depended most heavily on the advantages of being in Manhattan's Garment Center. The attractions of the Region have enabled it to remain far ahead of other apparel centers, but have not been powerful enough to offset the outward shifts. The Region's percentage share of the whole industry group has long been declining, and even its absolute employment now appears to have reached its peak.

In printing and publishing, certain kinds of activities flocked to New York as fast as the large-scale printing operations fled. Magazine publishers and book publishers found midtown Manhattan increasingly congenial; they moved there from other cities, or more often, they simply had their origins in Manhattan. The Region also experienced a steady accretion of miscellaneous printing which had to be near the customers who placed the orders and which grew because such customers steadily increased in New York. Among these customers were advertising agencies and the central offices of corporations. The Region's pulling power for publishing and for such printing as direct-mail advertising has been so consistently strong that, despite the exodus of much standardized printing, the Region's percentage share of national employment in printing and publishing as a whole is as high today as it was in 1900—a remarkable fact in view of the westward march of population and income.

In electronics, the operations with the closest affinity to the Region were those conducted by the firms which live with uncertainty—the new firms, the firms making new and unsettled products (which are always emerging in that industry), and most of all the firms making unstandardized products, such as hi-fi sub-assemblies, special-purpose tubes, and electronic systems for missiles, planes, and naval vessels. The strong points of the Region have kept it a major electronics center with a fast-growing employment, but its share of the industry has gradually decreased.

Again, it is not hard to find common themes in our three industry groups. In all three, the activities which took most readily to the Region tended to be activities resulting in products whose characteristics could not be accurately anticipated—products, that is, which were unstandardized.

It was not only that the final products were relatively unpredictable; the inputs were, too. The making of the products required materials and labor and services that were not easily determined in advance, either as to their exact nature or their amounts. Dresses subject to style changes, for example, have changing inputs of design, fabrics, dyes, tailor-work, buttons, and finishing touches. The publishing of a magazine (though not the printing stage) requires ideas, articles, fiction, pictures, and all sorts of special typographical and artistic work for advertisements—a diverse and changing bundle of inputs that cannot be purchased like bricks. The building of an electronic navigational system for a supersonic bomber requires engineering, skilled labor, and a multitude of special parts that cannot be bought over any counter.

Uncertainty over the product and the inputs causes a manufacturing establishment to need flexibility. Uncertainty over the *volume* of output has a similar effect. Establishments for which uncertainty is a normal part of life cannot generally achieve a high degree of self-sufficiency. Rubbing elbows with others of their kind and with ancillary firms that exist to serve them, they can satisfy their variable wants by drawing upon common pools of space, labor, materials, and services. In more concise language, they can take advantage of external economies.

The economies are external in the sense that the firm obtains them from outsiders, and they are economies in the sense that the firm can satisfy its variable or part-time needs in this manner more cheaply than it could satisfy them from within. The outsider, in turn, can afford to cater to the firm's fractional needs because he also caters to many other firms. The external economy may derive from an electrician or a sewing machine repairman or a free-lance photographer, responding to the call of a firm which does not need him full-time. It may derive from a manufacturing establishment doing specialized contract work such as embroidery, typesetting, photoengraving, or the making of unique electronic components. It may come from a supplier of buttons, fabrics, thread, or paper, able to make fast delivery so that the manufacturer does not have to keep large numbers of things in stock. It may grow out of a testing laboratory, a technical library, a convenient cluster of hotels to accommodate visiting buyers, or a freight forwarder pooling the small shipments of small firms into carload lots. It may be based on the presence of manufacturing space in small, variable, rentable pieces. It may even grow out of a revolving supply of specialized labor, such as garment workers accustomed to seasonal cycles, printers, staff writers, editors, or electronic engineers. Such a supply enables a firm to pick up employees quickly and let them go with equal suddenness, and makes it unnecessary to maintain a stable force of workers for an unstable demand.

Thus it is obvious that external economies reduce the cost of doing business just as labor and transport do. Indeed there is no real line of demarcation, for an external economy can be based on a labor supply or a railroad. But the practical distinction for our purposes is that in the case of labor or transport costs, one locality can be more easily compared with another in numerical terms like wage rates and freight rates, while in the case of external economies—as used here—the emphasis is on their *availability*.

One of the chief attractions of the New York Metropolitan Region for manufacturing has been its impressive array of facilities which generate external economies. Whether these facilities came into being to serve particular industries or whether they simply grew up as inevitable ingredients of urban life and served many industries, their presence has acted as a magnet to attract and hold the kinds of manufacturing that needed them the most.

The Region possessed another attraction, first cousin to external economies but having an individuality of its own. The metropolitan area, most of all the roaring city at its center, could offer the means of reducing the cost and time of personal communication. Fast face-to-face transaction of affairs is an essential condition for much that goes on in a city. It goes far toward explaining the city itself. Manufacturing taken as a whole does not require as much personal communication as some other economic pursuits—for example, trade or banking—but *some* phases of manufacturing require it urgently. And those that do are just the kinds of activities which, in our three industry groups, have felt most strongly the magnetism of New York.

The creation and production of high-style dresses, the preparation of text, pictures, and advertisements for a weekly magazine, the printing of miscellaneous materials with early deadlines, the building of electronic equipment for airplanes and rockets—such activities require frequent consultation among the people and establishments that have some part in the affair. The less standardized the product the more necessary it is to consult, instruct, solicit, question, buy, sell, inspect, reject, argue, alter, check and double-check along the way. What is more, all this must generally be performed at a breathless pace. The dress firm tries to get ahead of its competitors with a new style, and if a firm comes up with a successful number, other firms scramble to copy it. The staff of a magazine strains to conquer the deadline and the opposition. The maker of a new or changing electronic product hastens to get his version ready for sale to the public or the government before somebody else beats him to it.

Personal communication is not confined to those who have a contribution to make to the product as it takes shape. Some of the most valuable communication of all is that which enables a firm to stay knowledgeable in its field. A firm needs access to ideas and information for use in its next model or edition, and access to trade gossip on what is going on in the industry. This is especially true of manufacturing in which the product continually changes. The Manhattan luncheon table as a provider of ideas is a far more valuable cost-saver to some firms than the lower cost of labor in some other city would be.

Thus the principal advantages of the New York Metropolitan Region for the industries analyzed in our case studies are external economies and the means of reducing the costs of personal communication. These advantages, it is clear, determine the location of some manufacturing activities as surely as the lower cost of labor or transport determines it for others. Unstandardized activities, by clustering close together, using one another, and living off the motley urban environment, can operate more economically than they could if scattered in small cities and open fields. This observation is a commentary on the role of big cities in general, but it is especially apt as a commentary on the biggest city of all, including the suburbs in its economic orbit.

When industries change

So far we have been discussing the locational factors in which one geographical area has an advantage over another. These competitive factors—labor, transport, external economies, and personal communication—have the effect of causing any given slice of manufacturing to be tugged simultaneously from many directions. And there are situations in which still other competitive factors may govern, examples being the cost of space, the cost of taxes, the availability of wood or water or glass, the cleanliness of the air.

But we cannot simply leave it at that, for our case studies make it plain that the shifts of manufacturing are touched off not only by the conditions to be found in geographical areas but also by the dynamic technological changes that take place in manufacturing operations. These changes cause individual firms or pieces of manufacturing activity to alter their requirements and therefore to respond differently to the relative attractions of different locations.

Products tend to change in the degree of their standardization. Firms tend to expand or contract, thereby reducing or increasing their need of external economies. Industries tend to reconstitute themselves on a new basis to meet new conditions. And manufacturing activities also are caught up by technological advances in the economy as a whole—for example, the coming of the motor truck and the highway, of air freight, and of the teletype machine.

Changes of these kinds constitute a second dimension of locational factors. The shifts that have occurred in our three industry groups must be viewed in this dimension as well as in the dimension of geographical differentials.

When a product became more standardized, as radio sets did when their technology settled down, firms which assembled them could afford to make larger commitments for space and equipment. Producing on a bigger scale, they could reduce their dependence on external economies by replacing them with internal economies; for example, if an establishment now had enough repair work to keep a repairman busy all the time, the management was free to choose whatever arrangement was cheapest. Being able to anticipate better the form and volume of the output, firms could divide the work into simple operations which could be performed by unskilled labor. All these developments pulled together in favor of locations where cheap labor was available and space was plentiful.

But even if a product did not become more standardized, some of the stages of its manufacture might be more standardized than others. And if it were technically feasible to separate a more standardized stage from a less standardized one, there might be cases in which costs could be reduced by placing the stages at different locations. This kind of separation, this "disintegration" of functions, shows up repeatedly in our case studies in connection with manufacturing shifts.

The making of women's clothing provides an illustration of the locational importance of the separation of functions. New York's main attractions were its special kind of labor pool, its unparalleled external economies of other kinds, and the opportunities it afforded for meeting, mingling, and communicating. But as time went on, garment manufacturers began feeling the pull from other areas where labor was cheaper and not so strongly organized. Even so, the other areas had to pull hard indeed to overcome the attraction of Manhattan's busy Garment Center. Their success in overcoming it has been due not only to the geographical cost differentials but also to certain other developments which acted as catalysts to facilitate the process of shifting to more economical locations. The industry learned how to produce certain garments—the more standardized ones—in a way that reduced the need for skilled tailors. That is, the labor was divided up among more workers, so that each worker performed only a relatively simple operation. Further, the industry learned how to effect a physical separation between the functions that most urgently needed fast personal communication or external economies and those that needed them less. Under the new system a firm could remain in the Garment Center but could contract part of the process out to other shops. And the rise of the truck and the highway made it unnecessary for these other shops to be in the same locality; instead they could be in low-cost areas anywhere within overnight trucking distance of the home base. Finally, another element entered the brew: a change in public tastes in favor of simpler, more casual garments, more easily produced by unskilled labor.

This is not the whole story of the outward movement of apparel production from New York, but it is enough to give an idea of the factors at work. Similarly, in the magazine industry, the migration of magazine printing to the Midwest was the result not only of lower transport and labor costs there but also of the improvements in inter-city communication and transportation (such as the teletype machine and the airplane) that made it feasible to separate the standardized printing stage from the publishing stage which had to remain in New York.

Past and future

It would appear, then, that in the three industry groups under scrutiny, the experience of the New York Metropolitan Region has been something like this:

When transportation cost has played a role in the movement of industry, it has generally worked to the disadvantage of the Region. The same has been true when labor cost has been a governing factor—especially the cost of unskilled labor. When skilled or professional labor was the prime need, the Region has shown more pulling power. The Region has exhibited a capacity to capture relatively fast-growing splinters of industry at an early stage of their growth. And when either uncertainty of product or small size of establishments has put a heavy premium on external economies, or when fast personal communication has been an indispensable condition of successful operations, the Region has performed at its best. Thus, something like an endless process of adaptation has been taking place, with some manufacturing activities continually thriving and growing in the special environment of the New York

Metropolitan Region and with other activities in the same industry groups continually breaking away piecemeal to pursue the advantages of other locations.

As for the future, each of the case studies ends with projections of employment in the Region, reaching to 1965, 1975, and 1985. Though we call these "projections," the word has a restricted meaning in the context. An analyst focusing on the forces at work in a particular industry may be in a better position than anyone else to say how much of the industry is likely to settle in an area like the New York Metropolitan Region, but he cannot be all-seeing. Apart from the murkiness of economic forecasting in general, the industry analyst cannot estimate how all the other activities that go on in the Region may be behaving during the period of his projection. Hence, he runs the risk that the labor supply on which he had counted will be siphoned off for other uses, or that land will grow scarce, or that events beyond his ken may raise taxes to intolerable levels.

It remains for the final volume in this series to take all these forces into account and to provide an integrated and internally consistent picture of the Region. This integration could portray a slightly different future for the three industry groups than the present volume affords. But as a synthesis of all that our three analysts knew and all that they were prepared to guess, the forecasts in this book serve a substantial purpose.

Even without the forecasts the case studies, with their insights into the Region's weaknesses and strengths, tell much that has a bearing on the future. Through such insights, those who hope to shape the course of the Region's development can gain a better sense of the forces with which they must deal.

WOMEN'S AND CHILDREN'S APPAREL

By

Roy B. Helfgott

Appendix A, Classification of Apparel Industries, begins on page 329.

Appendix B, Additional Apparel Statistics, begins on page 332.

Notes to Women's and Children's Apparel begin on page 333.

I

The Business of Garment-Making

Put the American woman on a subway train going from Pennsylvania Station to Times Square, and it will take her just 65 seconds to pass completely under the district that gives her the reputation of being the best-dressed woman in the world. Manhattan's famous Garment Center, as shown in Chart 1, is bounded approximately by 34th and 40th Streets, and by Sixth and Ninth Avenues, with Broadway slanting through the east portion. This small area, a place of trucks and bulky buildings, bolts of fabric, racks of dresses, drawing boards, sewing machines, showrooms, and innumerable people in a hurry, makes New York a fashion headquarters second to no other city, with the possible exception of Paris. Though the influential Parisian *haut couturier* provides many ideas, the clothes for American women are fashioned chiefly in the Garment Center. The firms there, plus those in the rest of New York City, account for about two-thirds of the sales of all women's and children's garments in the United States, though much of the actual production is done elsewhere.

The city's dominance in the women's apparel industries dates back almost to the advent of ready-to-wear clothing in the second half of the nineteenth century. In recent years, however, the dominance has been weakening, and doubts have arisen concerning New York's future as the fountainhead of women's apparel. Some see New York losing out as the fashion capital to Los Angeles, St. Louis, Dallas, and other major cities. Others see New York continuing as the great center for ideas and designs but not for the production of the garments.

Our purpose in this report is to describe what has been happening to the women's and children's apparel industries in New York and to indicate possible future trends. We are concerned not merely with the Garment Center but with the whole New York Metropolitan Region of 22 counties, for of course the Garment Center is not the sole scene of action; women's underwear firms are farther east and south in Manhattan, knit outerwear is centered in Brooklyn, and various types of garment production are found at many points of the Region.

Chart 1

The Garment Center and Nearby Areas of Midtown Manhattan

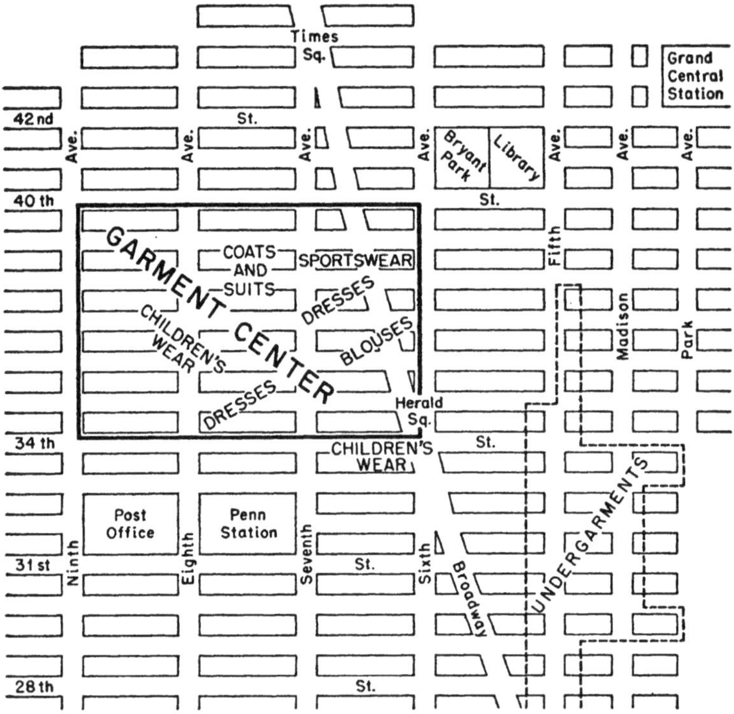

Names of garments show heavy concentrations of manufacturers and jobbers.

To accomplish our purpose, it will be necessary throughout the report to analyze the women's and children's apparel industries, their labor force, the demand for their products, and recent technological changes. The present chapter will contain a discussion of the business of making garments—a volatile and sometimes fantastic business whose peculiarities must be understood in order to make reasonable guesses as to the future location of its shops. In the other chapters we shall attempt to answer three main questions: (1) How the New York Metropolitan Region became the nation's garment headquarters; (2) Why it is declining now; (3) What its future will be like.

The report covers all women's and children's apparel except fur garments, footwear, headwear, and accessories. Most of the products fall into three broad classifications: women's and misses' outerwear, women's and children's undergarments, and children's outerwear. In addition it is appropriate to take a cursory look at knit outerwear, which, though officially classified as a textile industry, produces style garments that are often interchangeable with the products of the apparel industries. It thus behaves locationally like those industries and is heavily concentrated in New York. In all we are talking about twelve separate industries as classified by the Bureau of the Census. They can also be considered as one broad industry, for "industry" is a remarkably elastic term, and when we discuss their common characteristics we will not hesitate to use the singular. The twelve are listed in Table 1, and, for any reader who is interested, they are more fully described in Appendix A.

These industries hold extreme importance for the New York Metropolitan Region. They provide more than a quarter of a million jobs, or 15 per cent of the Region's total manufacturing employment. They are even more important to New York City itself, since they employ nearly 200,000 there—one out of every five manufacturing workers—and outrank all other comparable groups of industries in the metropolis. The magnitude of apparel-making in the city and the Region, moreover, attracts numerous auxiliary industries and services: for example, embroidery, pleating and stitching, buttons and bows and belts, the textile-design and fashion-design firms, the making of pins and needles, the repair of sewing machines, the administration of the union's affairs, the hostelries where out-of-town buyers hang their hats.

Omitting the supporting activities and considering only our twelve women's and children's apparel industries, we can illustrate the national importance of the Region by four simple percentages for 1954, when the federal government took a Census of Manufactures. In that year and in those industries, the Standard Metropolitan Area,* which includes five fewer counties than the New York Metropolitan Region as defined in our study, had 43 per cent of the nation's employees; 62 per cent of the establishments; 51 per cent of the "value added" by manufacture; 66 per cent of the value of sales.[1] The differences among these four figures tell a good deal about the women's and children's apparel industries—for example, that establishments in the New York Metropolitan Region are smaller in size than elsewhere, and that the Region has a greater share of national sales than of national production. These significant facts will be discussed in their proper turn.

✓ Business organization

In today's business world, where an industry typically consists of a few giant firms having huge aggregations of capital, turning out similar products at identical prices, and making large outlays on research and development, the manufacture of apparel seems to be an anachronism. Here is an industry that still adheres to the nineteenth-century pattern of "free enterprise," meaning small, personally conducted firms engaged in a competitive market.

The key to success in the industry is not continued investment

* The Census Bureau calls this the "New York–Northeastern New Jersey Standard Metropolitan Area." In this book we will ordinarily shorten it to "New York Standard Metropolitan Area," or, in appropriate cases, simply to "the Standard Metropolitan Area," as above. The five counties which are in the New York Metropolitan Region but not in the Standard Metropolitan Area are Orange, Dutchess, and Putnam in New York State; Monmouth in New Jersey; and Fairfield in Connecticut.

in new plant and equipment, for the entire process of production is secondary to the prime needs of style creativity and merchandising—"one good style and you're set." Style differentiation and the rapid obsolescence of garments as one style replaces another in the hearts of consumers intensify the importance of the merchandising function. The saleability of a garment depends on a combination

Table 1 Women's and Children's Apparel Industries, United States, 1954

	Number of establishments	Number of employees (thousands)	Value of shipments (millions)
Total, all industries	15,207	600.1	$6,256.1
Women's and misses' outerwear, total	10,298	363.9	3,805.2
Blouses	1,245	43.0	375.3
Dresses, unit-priced	4,114	143.3	1,455.1
Dresses, dozen-priced	892	54.5	435.2
Coats, suits, skirts	3,204	96.0	1,261.3
Neckwear	133	1.7	26.3
Other outerwear	710	25.4	252.1
Women's and children's undergarments, total	1,843	112.2	1,164.8
Underwear, nightwear	1,352	73.4	770.0
Corsets, brassieres	491	38.8	394.8
Children's outerwear, total	1,983	77.5	741.9
Dresses	764	32.5	291.5
Coats	407	14.4	172.1
Other outerwear	812	30.6	278.2
Knit outerwear	1,083	46.4	544.1

Source: U.S. *1954 Census of Manufactures.*

of the professional ability of the firm's designer and the merchandising acumen of the firm's director.

Since production of the garment is secondary to its design and sale, a separation has taken place between production and the other operations. In point of fact this separation is very old. There is evidence in British as well as American history that it originated in the practices of merchant employers who gave work to journeymen tailors to be done in their homes. The coming of the factory brought a considerable integration of functions, but separation reappeared in the United States during the closing decades of the nineteenth century when Jewish immigrants arrived in large numbers and manufacturers began to farm out their excess production to contractors who organized the immigrant labor. Separation got a further boost in the 1920's with the rise of the jobber, an important Garment Center character who is neither a manufacturer nor a contractor.

Nowadays, as a result of the separation of functions in the industry, when one speaks of the employer he must specify which of three types he has in mind. The three types are:

1. *Manufacturer,* who has an "inside" shop—that is to say, besides buying the cloth and selling the final product he runs his own production plant, in which his workers cut and sew the material into apparel.

2. *Jobber,* who buys the raw materials, designs the garments, and later sells them, but does not actually manufacture them. (In knit outerwear such a man is called a "converter" rather than a jobber.) The jobber has a showroom, and usually a cutting room, but sends the fabric to "outside" shops to be sewed into apparel.

3. *Contractor,* the man who runs the "outside" shop. He has the establishment where the fabric is sewed into garments, and hires the workers, and is, in effect, a labor contractor. He produces to specification, never takes title to the goods, and is in no way involved with the marketing of the product.*

The jobbing-contracting relationship has become a solidly established method of operation because it conveys two strong advantages. First, it permits specialization; the jobber concentrates on the merchandising while the contractor is a production expert. Sec-

* The boundary lines between the three categories are perhaps not quite so clear-cut as these definitions suggest. Some establishments are on the border line, difficult to classify. And in some cases what is ostensibly an independent contractor shop is in reality a subsidiary, owned by a jobber or manufacturer.

ond, it adds flexibility to an industry where uncertainty reigns; it provides a means of keeping down the unused plant capacity of all concerned.

This second advantage offers a key to understanding the business organization in the women's and children's apparel industries. Although aggregate demand for their products is predictable, the style factor prevents any single firm from knowing what its volume of sales for the season will be; thus the use of contracting relieves each firm of the need to maintain factories large enough to fill maximum orders. The jobber can find additional contractors to produce for him when his orders rise, and the contractor, in turn, is not dependent upon orders from only one jobber, but may turn to others to keep his plant occupied. Contracting even helps the manufacturers who operate their own plants, because those plants need only be large enough to handle normal demand, and peak demand can be contracted out.

The prevalence of contracting in the women's and children's apparel industries is made evident by Table 2. Only 38 per cent of the 15,000-odd establishments in the United States are the "inside" shops of manufacturers, and only 45 per cent of the total employment is to be found there. The contractors with their "outside" shops have 46 per cent of the establishments and 46 per cent of the employees. The jobbers have 16 per cent of the establishments and 9 per cent of the employees. There are variations by industries, as the table shows.

Contracting is used most in women's outerwear, and least in undergarments. Children's outerwear falls between the two. In only one industry, corsets and brassieres, are as many as two-thirds of the establishments those of manufacturers. This industry, producing the most standardized of all the types of women's and children's apparel, differs in many respects from the rest of the garment industries. At the other end of the scale are blouses and unit-priced dresses, which are subject to frequent style changes. In those two industries only about one-quarter of the establishments are those of manufacturers.

The incidence of contracting varies not only by industry but also by location. New York is the center of contracting, while most apparel centers in other parts of the country rely on inside shop production.[2] In fact, inside shop production in the style items—unit-priced dresses and coats and suits—is declining among New York-

Table 2 Percentage Distribution of Manufacturers, Jobbers, and Contractors, United States, 1954

	Establishments			Employees		
	Manufacturers	Jobbers	Contractors	Manufacturers	Jobbers	Contractors
Total, all industries	38%	16%	46%	45%	9%	46%
Women's and misses' outerwear	33	16	51	36	10	53
Blouses	23	18	59	21	10	68
Dresses, unit-priced	27	16	57	29	12	59
Dresses, dozen-priced	49	8	43	49	7	44
Coats, suits, skirts	36	18	46	43	10	47
Neckwear	59	17	24	58	19	23
Other outerwear	39	15	46	47	8	45
Women's and children's undergarments	57	13	31	62	6	32
Underwear, nightwear	53	15	32	63	5	32
Corsets, brassieres	67	6	27	60	8	32
Children's outerwear	44	17	39	47	11	42
Dresses	40	16	43	36	14	50
Coats	39	20	41	47	10	43
Other outerwear	50	15	35	59	8	33
Knit outerwear	51	14	36	65	8	27

Note: Because of rounding, not every trio of percentages will add exactly to 100.

Source: U.S. *1954 Census of Manufactures.*

based firms. This has locational implications, because most of the manufacturers' establishments which fold up are in Manhattan.

Chart 2 shows how the three leading states in women's and children's apparel, New York, Pennsylvania, and New Jersey, compare with one another and the rest of the country in the number of people employed by manufacturers, jobbers, and contractors.

Chart 2

Three Leading States' Shares of Employment in Women's and Children's Apparel Industries, by Type of Establishment, 1954

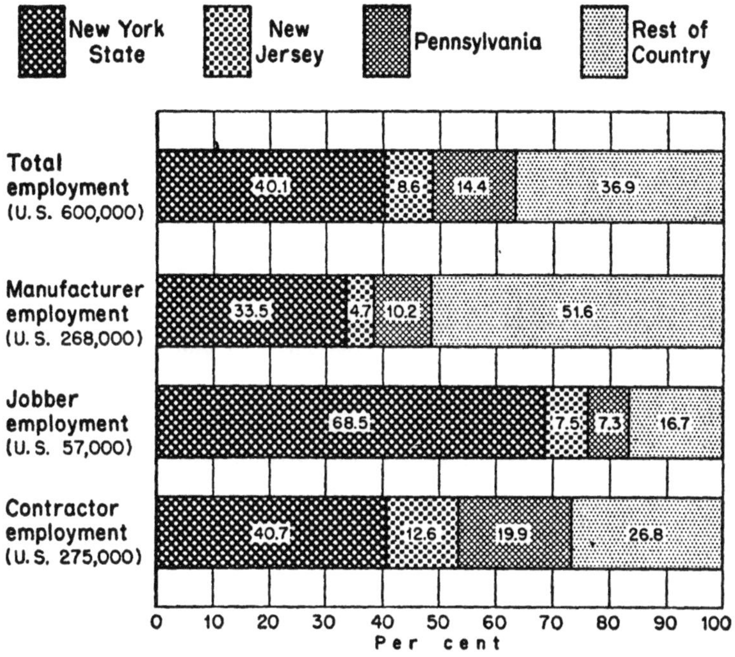

Source: U.S. *1954 Census of Manufactures.*

Notice that their national dominance is far more impressive in jobber and contractor employment than in manufacturer employment. Notice also that New York State, which in garment-making means chiefly New York City and adjacent counties, has more than two-thirds of all jobber employment but not nearly so much of the other kinds.

✓ Ease of entry and exit

The women's and children's apparel industries remain an area of American economic life in which an individual with some know-how, a little capital, and plenty of fortitude can enter business with reasonable hopes of success. Profit rates are not high, and in the most recent year for which net profits are known, they were only 1.34 per cent of sales. In that year, 1953, more firms had losses than profits.[3] Despite this, there are always people who are willing to take the gamble. Fortunes can be made overnight—all it takes is a single style or "number" that captures the fancy of the buyers.

Capital requirements vary product by product; for example, a firm intending to produce undergarments in its own large factory must invest a great deal more than a firm entering a style industry. But a dress or blouse firm in New York may enter business with as little capital as $15,000, though normally it is closer to $25,000. Some dress firms may start with a little more, but only occasionally does one start with as much as $100,000. The low capital requirements are made possible by the fact that a firm need not engage in production; it can use contractors instead. Outside New York, in places such as St. Louis, Dallas, and Miami, where inside manufacturing is more important, the capital requirements are greater.

The manufacturing or jobbing firm may be launched by two persons, one of whom has been a salesman and the other a cutter or patternmaker. Since selling is such an important part of operations, and since it is most difficult to hire a good salesman (they strike out on their own too fast), it is essential to have one proprietor with good retail contacts. Most New York houses do their own cutting in order to keep control of the piece goods; therefore the cutter is also needed as a principal in the business to serve as a production expert. In the better price lines, where houses do their own manufacturing and where style creativity is a paramount consideration, the principals of the firm may include not only a production chief but also a designer.

The typical firm is a closed corporation, all the capital being contributed by its principals. In effect, the firms are partnerships, but the corporate form of business organization is used in order to obtain the advantage of limited liability. In apparel there has ordinarily been no absentee ownership by stockholders; the head of the firm has run it. Of late, however, a possible beginning of absentee ownership is detectable in those industries, such as dozen-priced dresses, which produce the most standardized garments.

The partners enter business with great expectations, which are seldom realized. Turnover is rapid. Some firms survive no more than a season or two. A few go on to become established names in the trade, but the average life span is in the neighborhood of seven years. In 1950, only 13 per cent of the firms (that is, manufacturers and jobbers) in the women's and children's outerwear industries had been in business more than twenty years. In undergarments the percentage was 25.[4]

Liquidations usually mean that new firms arise from the old ones. To some extent, the turnover is a game of musical chairs: Old firms liquidate; the principals then join principals from other firms in new business combinations. Even those who are forced out of business and return to the ranks of the employees are often only biding their time before having another fling. Thus, although firms come and go, the individuals stay around much longer.

A very high amount of business is done relative to the initial capitalization. Dress firms, on the average, turn over their capital seven or eight times a year. Credit is necessary as a permanent feature of the apparel industries, and the typical businessman lives on credit. One reason for this is the seasonal character of apparel-making. For example, the jobber or manufacturer buys piece goods in December—on a 60-day or 70-day credit either extended by the textile mill or through intermediaries—and turns them into garments between January and June. Early in the period no money is coming in; so he is further dependent upon credit for working capital needs such as meeting payrolls. For such expenses he borrows from commercial banks on 90-day notes, which are corporate notes but are personally endorsed by the principals of the firm. Since firms are usually prompt in their payments, the banks are liberal in their credit policies, and a firm's loans often equal its working capital.

The firm with a good reputation is in a fairly good position credit-wise, since it typically receives 70-day credit from the textile mills and 90-day loans from the commercial banks, but sells to retailers on terms of 30 days; and although it does not always collect within the 30-day period, it is still ahead of the game. Even shaky firms can get loans by putting up their accounts receivable as collateral and by paying higher interest.

There are often bad-debt losses when firms fold up. Though the creditors are generally happy to receive 50 cents on the dollar, the principals are usually back in business the very next season, having rounded up additional capital from relatives and friends. Most liquidations do not reach the courts; the entire matter is handled —without bankruptcy petitions—by the New York Credit Men's Adjustment Bureau, Inc., an affiliate of the New York chapter of the National Association of Credit Men. The work of the Bureau is a locational factor of some importance to apparel firms, and so is the presence of extensive credit facilities in New York. Both the Bureau and the wide choice of lending institutions, including commercial banks and "factors" (textile bankers), can be viewed as services available to a firm in the nation's first city that are not so readily available elsewhere.

Size of the Establishments

The nation's 15,000-odd establishments in the women's and children's apparel industries in 1954 had only 39 employees each, on the average. The average varies in accordance with the type of establishment (whether manufacturer, jobber, or contractor), the industry, and the location.

Contractors employed an average of 39 persons—the same as all types—with manufacturers employing 46 and jobbers 24.

As for the variations among industries, the average number of employees per establishment (leaving aside the distinctions between manufacturer, jobber, and contractor) ranged from 71 in corset-and-brassiere shops to 13 in neckwear shops. Another way of showing these differences is to give the percentage distribution of employees by size of establishment, and this is done for all twelve industries in Table 3.

The differences are linked quite closely to the nature of the garment. The more standardized the product, the greater the division

Table 3 Percentage Distribution of Employees by Size of Establishment, United States, 1954

	Employee-size classes				
	1–19	20–49	50–99	100–249	250 or more
Total, all industries	10.5%	26.6%	24.4%	22.7%	15.8%
Women's and misses' outerwear, total	12.0	31.8	26.3	19.1	10.9
Blouses	12.3	33.3	29.0	20.8	4.6
Dresses, unit-priced	11.4	38.7	28.7	15.8	5.5
Dresses, dozen-priced	5.2	13.7	20.6	31.4	29.2
Coats, suits, skirts	16.1	32.0	25.6	17.5	8.8
Neckwear	42.1	30.4	27.4 [a]	....	0
Other women's outerwear	11.4	28.3	23.7	15.6	21.0
Women's and children's undergarments, total	5.8	14.4	18.5	30.7	30.6
Underwear, nightwear	7.1	16.7	20.7	28.0	27.5
Corsets, brassieres	3.4	10.0	14.2	35.8	36.5
Children's outerwear, total	10.4	24.7	25.9	26.3	12.7
Dresses	8.7	24.6	27.0	25.3	14.5
Coats	12.7	29.0	27.5	30.7 [a]	....
Other outerwear	11.2	22.8	24.1	25.2	16.7
Knit outerwear	10.2	19.4	21.0	25.4	24.1

Note: Because of rounding, not every line will add horizontally to exactly 100 per cent.

[a] Includes next class as well. This will cause insignificant errors in some of the "totals."

Source: U.S. *1954 Census of Manufactures.*

of labor in its manufacture, and hence, the larger the establishment. The most standardized garments for women are the corsets, brassieres, and allied items; that is, they are the least subject to rapid and violent style changes, and therefore each style can be manufactured in longer runs. The industry producing these garments also has the greatest proportion of its employees in large establishments, with over 70 per cent in establishments of 100 or more employees. The "dozen-priced" dress industry, producing inexpensive household apparel sold by the dozen, has 60 per cent of its employees in establishments of 100 or more. Next comes the underwear-and-nightwear industry, in which 55 per cent of the employees are in the 100-or-over category.

At the other extreme from corsets is the women's neckwear industry where almost three-quarters of the employees work in shops of less than 50 employees. Similarly, in the fashion-oriented industries—blouses, unit-priced dresses, and suits-coats-skirts—three-quarters or more of all employees are in establishments of under 100, and about half in establishments of under 50.

There is also a high correlation, reflecting the link with standardized product, between the size of establishment and the concentration of industry in and near New York. This correlation shows up in Chart 3. The industries with the smallest shops (and incidentally with the least standardized products) are also the ones that cluster most heavily in the 17 counties of the New York Standard Metropolitan Area. The correlation also holds true *within* the industries; in fact, for every one of the women's and children's apparel industries we find that in the states where contractors work mainly for New York jobbers, the shops of these contractors increase in size with the distance from New York City. The unit-priced dress industry will serve as an illustration. In that industry the average-size contractor establishment in New York State had 30 employees; in New Jersey, 36; Connecticut 55; Pennsylvania 59; Rhode Island 68; and Massachusetts 69.[5]

Not only is the typical apparel *establishment* small, but so is the typical *firm*. One of the largest dress firms, Jonathan Logan, Inc., with more than 40 establishments, has a reported annual sales volume of $36,000,000,[6] which is less than 3 per cent of the total sales of unit-priced dresses in the United States. Even 3 per cent, however, is an overstatement because this company, though its chief product is unit-priced dresses, also manufactures blouses, slacks, and still other items.

Chart 3

Relation Between Size of Establishment and Concentration in New York Standard Metropolitan Area, 1954

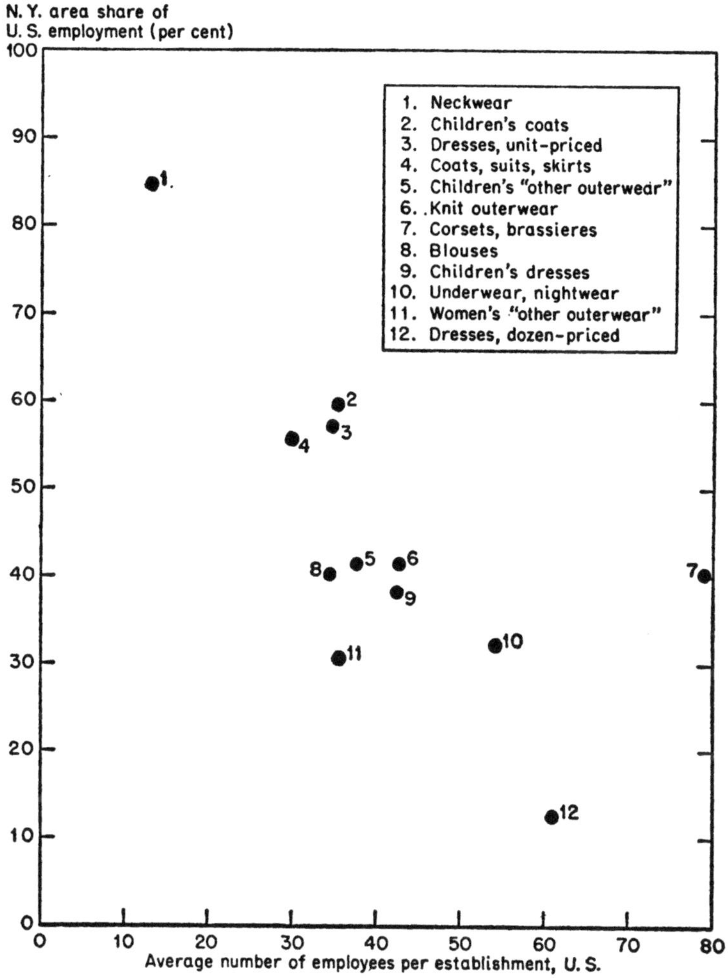

Source: U.S. *1954 Census of Manufactures.*

The price competitiveness of the industry is illustrated by the fact that very few producers of women's and children's apparel engage in national advertising. Only the corset-and-brassiere industry, whose products are fairly standardized, engages in consumer advertising to any great extent.

Production methods and technology

The small size of establishments and the low capitalization of firms are due to the fact that economies of scale are definitely limited in the women's and children's apparel industries. Technological advance is severely inhibited by the nature of the production process, of the materials worked on, and of the demand for the products. An essential element of fashion is individuality, and a woman "would rather be caught dead" than seen at a party in the identical dress worn by another woman. As a result, major branches of the women's and children's apparel industries must be geared to quick changes in style. The small plant with its inherent flexibility can more easily adapt to such changes.[7]

Small-scale retail ordering of diversified styles leads to small-scale methods of production. This helps to keep the industry technologically backward, because:

> Under these conditions the energies of manufacturers, even those with indoor [inside] factories, tend to be concentrated on marketing, that is, on increasing seasonal orders by offering a wider and more up-to-the-minute variety of styles each season, rather than on improvements in production involving the purchase of new machines and the reorganization of production methods to increase output per operative.[8]

Originally, when the market for women's clothing was restricted to the small group of wealthy people, there was no division of labor, and a custom tailor sewed the entire garment by hand. With the invention of the sewing machine and coming of garment factories, the production process was broken down into the constituent crafts of cutting, machine sewing, hand sewing, pressing, and examining. Under this system, known as tailor work (single hand), one machine operator sews out the bulk of the garment. Such workers are skilled craftsmen, who must be familiar with all parts of the garment. This system is most efficient for the small shop, which works on small runs of many styles. Little inventory in process is required and goods can be produced on short notice. The system has been associated mainly with the older trades—coats and suits and dresses—and with the immigrant Jewish and Italian workers in New York and other large cities. Tailor work remains prevalent in the older centers of production, particularly in the higher price lines, which require small runs of many styles.

Since fashion is much less prevalent in men's clothing than in women's, the factories in the men's clothing industry started to discard tailor work before the end of the nineteenth century. They took up a new system known as section work. In this system the labor is divided. The machine sewing of the garment is broken down into a number of operations, the work being passed from operator to operator by hand. A shop needs only a core of skilled workers for the more intricate operations, while most of the workers perform the more routine and simple operations which require little training. When the method of production becomes more complicated, as in section work, the need for efficient management increases, the overhead cost rises, and the plant loses flexibility in its operations. Elaborate systems can become white elephants if the volume of work fluctuates greatly, because large segments of the factory and labor force must be used regardless of the output. In those industries subject to frequent style changes there is less opportunity to introduce complicated systems.

Because of the style factor, section work came late to the women's apparel industry. Today, however, it is dominant in those branches, such as corsets and brassieres, dozen-priced dresses, underwear and nightwear, which produce fairly standardized garments. And it has even spread into the popular price lines of the style industries—coats and suits and unit-priced dresses. New, large section-system plants have grown up on the peripheries of the major garment centers or in new centers of production.

In addition to a division of labor, there has been modest technological progress. The desire to exploit the potential urban market led to the mechanization of the clothing industry in the middle of the nineteenth century.[9] Mechanization was made possible by the invention of the sewing machine by Elias Howe in 1846 and the improved model put on the market by Isaac Singer four years later. A century ago the federal government reported, "A large manufacturer in New York who commenced business in 1849 was the first to introduce sewing machines in the business. . . ."[10] Other innovations followed, including the use of steam power and later electric power. Buttonhole, pressing, and cutting machines were invented, and so were special-purpose sewing machines. To this day, however, the single-stitch sewing machine remains the basic machine in the industry, though time has brought improvements in it. The most important improvement was the introduction of electric power; the machine of a century ago permitted the sewing of 800 to 900 stitches a minute, but with electricity, 2,000 to 4,000 could be sewed, and with today's late models, 5,000.

The human factor still holds its crucial importance in apparel, for the manufacture of women's wear remains essentially a handicraft industry tooled by power machinery.[11] Fast workers can turn out more with older machines than slow workers with new machines. Better machines, of course, permit fast workers to turn out still more. Today, the operator has at his disposal special machines and attachments to do tacking, felling (sewing strong seams), basting, and other tasks. As the division of labor is intensified, the incentive to develop special-purpose machines is increased.

Productivity

Despite the coming of section work and the introduction of numerous special-purpose sewing machines, there is little evidence of any great increase in productivity in women's and children's apparel. Indeed, the available evidence seems to indicate that production per man-hour in these industries has been decreasing, except for moderate gains in the corset-and-brassiere industry and more substantial gains in knit outerwear. Corset shops saw an annual decline of 0.9 per cent in the number of wage earners per unit of product between 1927 and 1937. Knit outerwear mills saw an annual decline of 2.8 per cent from 1923 to 1939. But the same economist who reported these productivity gains found none at all in the other industries.[12]

There are few recorded attempts to measure apparel productivity, and it cannot even be stated unequivocally that the division of labor increases output per worker, but under the optimum conditions of a large volume of production of a few basic styles, section work would seem to have the edge.

The refinements made in the sewing machine and improvements in the quality of management, particularly with regard to materials handling, plant layout, industrial lighting, office procedures, and stock and inventory control, are the basis for a contention that productivity is increasing in the apparel industries.[13] Some modest gains have arisen from these improvements, no doubt. But larger savings in labor costs have grown out of the fact that the garment specifications have been simplified. Today's garment requires much less work than that of fifty years ago; and this simplification of product has brought a saving in labor not to be confused with increased productivity.

There are some who foresee a real technological revolution in the clothing industry. Clothing, it is claimed, might be made of paper, the patterns die-cut, and the parts fused electronically. If this were to take place, the apparel industry as presently constituted would disappear, and large firms with substantial capital investments would become dominant. In terms of location, the industry would no longer be dependent upon New York, and firms would probably be inclined to place factories near their consumer markets, since transportation cost would rise in relation to the value of the product.

Such a development, of course, would undermine our analysis of the future of New York's women's and children's apparel industries, but there is every reason to doubt its occurrence.

For one thing, the technological adaptability of apparel manufacturing to capital-intensive mass-production operations is extremely questionable. An exception might be the production of brassieres, in which die presses have been introduced to replace ordinary cutting by electric knives. Such a thing was possible in brassiere plants because the parts of the garment are small and the production runs are very long. Even for this remarkably standardized product, however, attempts to mold foam rubber brassieres by heat, instead of sewing the pieces together, were abandoned as unsuccessful.

A second and equally important reason for doubting the likelihood of revolutionary technological change in the women's apparel industries is that women's garments then would no longer be fashion items. The economic utilization of mass-production techniques would require the output of thousands, if not tens or hundreds of thousands, of the same style. While this conceivably would be acceptable to the consumer in the case of coveralls for working in the garden or uniforms for waitresses, it seems most unlikely that many American women will soon be prepared to be seen in public wearing the same garment as everyone else.

When one discounts the likelihood of revolutionary changes he is justified in assuming only very modest future increase in productivity in most of the women's and children's apparel industries. Makers of the more standardized products, such as household apparel and undergarments, should experience moderate productivity increases, as some of them have in the past. But most of the industries will probably remain in a state of almost constant technology, and, this being so, their future manpower requirements will be dependent primarily upon demand for the products. The fascinating question of demand will be examined in later chapters.

Seasonality

A towering fact in the economics of the women's apparel industries is that they are seasonal. This fact, already hinted at in our discussion of credit facilities, has been an important one in terms of location, for seasonality has forced firms to locate in major centers having adequate reserves of skilled labor which can be tapped on short notice. Seasonality is especially important to New York, which has the largest reserve of skilled labor. Seasonality imposes a social cost on the community, in that it requires the payment of large sums of unemployment compensation, but it also brings to the community a great amount of employment that would otherwise go elsewhere.

The basic cause of seasonal fluctuations is the buying habits of the public. These habits are related to changes in the weather. Fashion also plays an important role; the more standardized and less subject to the whims of fashion a garment is, the less its production is seasonal, for the demand is steadier, and there is the possibility of producing for inventory. With high-fashion merchandise this possibility does not exist. Seasonality thus varies from industry to industry within the apparel group. In New York City in 1956, the number of employees varied by the following percentages between the low and high month of the year:[14]

Coats, suits, skirts	132.3
Dresses, unit-priced	37.1
Underwear, nightwear	27.8
Knitting mills	14.1

The coat, suit, and skirt industry, where employment in the highest month was 132 per cent greater than it was in the lowest month, is the most seasonal of all. It is also a style-oriented industry, most subject to climatic changes, and it is heavily concentrated in the Garment Center. The industry has basically two seasons a year—spring and fall; work on spring garments generally gets underway in November or December and continues until Easter Sunday, after which most workers are laid off until June, when work on fall and winter garments starts; the interruption between fall and spring work, however, is much shorter.

New York attracts the most seasonal elements in the apparel industry because it provides the greatest ease for filling rush orders.[15] Moreover, even within a given industry, establishments in New

York have wider fluctuations in employment than establishments elsewhere. For example, in the unit-priced dress industry, fluctuations are greater in New York City than in Pennsylvania, just as garments produced in New York tend to be more stylized and higher-priced than those produced in Pennsylvania.[16]

The fact that New York City is more seasonal in employment than the nation holds true for every women's and children's apparel industry. In 1956, New York City's range between the low and high months for women's outerwear was 36.9 per cent as against only 16.7 per cent for the nation; for undergarments it was 18.5 per cent as against 10.5 per cent; and for children's outerwear it was 28.1 per cent as against 10.2 per cent.[17]

Marketing, purchasing, and pricing

Before each season, in the establishments of the jobbers and manufacturers, designers are busy working up styles. Samples of the styles most likely to succeed are then made up for display during "market-opening week." When the big week arrives, the buyers for retail stores in all parts of the country converge on the showrooms of the Garment Center to see how the samples look on live models. After shopping the market, buyers begin to place orders. And as soon as the orders begin to come in, the apparel firm must take steps with textile houses to ensure its supply of materials. The purchase of fabric and the pricing of the garments are closely tied in with the marketing operation and will be discussed presently, but first let us look appraisingly at that Very Important Person, the Buyer.

Success for the apparel firm depends upon getting a "hot number" which will result in enough re-orders during the season to create a profit. The retail buyer's judgment is of crucial importance in the beginning of the season, because his reaction to the firm's merchandise spells the difference between success and failure. The buyer, in turn, must be wary lest his store be caught with a large inventory of garments which cannot be sold at existing prices.

Given the important role of the buyers, the industry caters to them. Entertainment of buyers is a cost of doing business. Free theater tickets, baseball game admissions, and Christmas gifts are customary. Even financial kickbacks have not been unknown, and a few years ago some buyers were sent to prison for accepting them. There is no precise way of measuring the economic impact of favors to buyers, but one thing certain is that Manhattan's hotels, theaters, and night clubs receive a substantial portion of their business in this manner.

To the retailers, New York's dominance as a sales center of women's and children's apparel has great advantages; they can save money by concentrating all their buying in one center.[18] Air travel enables a buyer not only to attend market openings but also to hop into New York every couple of weeks to keep abreast of the latest fashion doings.

But these visitors do not by any means do all the buying of women's and children's apparel. The smaller stores, unable to maintain buyers of their own, or unable to send their buyers to New York very often, can still enjoy the benefits of New York's sales dominance by purchasing through resident buying offices located in the city. The resident buyers not only do the purchasing for small stores but also keep larger stores continually informed on market developments.

The magnitude of the orders determines how much fabric the apparel firm must purchase from textile houses. But much fabric is also purchased in advance. In purchasing materials, uncertainty is prevalent, and judgment and plain luck may be the determinants of success. The apparel producer must be careful not to get caught with large inventories of fabric that he will not be able to use or so short of needed materials that he will not be able to fill orders.

The more a firm is willing to gamble, the better its opportunity to obtain the fabric it needs at lower prices. The situation is similar to that in the fur industry, where a purchase of skins at a low price can overcome almost any conceivable disadvantage in production costs, whereas a number of poorly timed purchases can, and sometimes does, mean the downfall of firms that are among the most efficient in the industry.[19]

With retailers' orders in hand and fabric in the house, the apparel firm begins to cut the garments with an electric cutting knife, perhaps as many as 100 or 200 at once. The cut parts are assembled into bundles to be sewed either in the firm's own plant, or, more usually, to be shipped by truck to the factory of a contractor, where they are sewed and pressed. If a contractor is used, the finished garments are returned by truck to the original establishment for shipment to the retailers.

The price the retailer pays for a garment has of course been set before the buyer sees it in the showroom. In setting the price, the jobber or manufacturer calculates the fabric and labor costs of the garment, and then adds his mark-up. In New York, mark-ups are generally lower than in other centers. In determining his mark-up, the jobber or manufacturer must figure in returns and end-of-season close-outs in order to arrive at gross profits. Such close-outs, in which he dumps merchandise at a loss, are inevitable in a fashion trade where one cannot accumulate inventory lest it get out of style and lose its value. Even piece goods cannot be saved unless they are solid-color staples, which can be used a year hence. Therefore, if one has over-ordered his fabrics, it is often better to make them up into garments to sell at a slight loss than to lose all on the fabrics. Dress houses produce against orders, with a three-to-four-week delivery, but in the early part of the season they will usually cut their cloth 50 per cent above orders in anticipation of re-orders—the promise of an early delivery date may make the sale —or on the assumption that their salesmen on the road will be able to push the extra garments. When re-orders fail to materialize, close-outs become necessary.

Apparel houses often give volume discounts on large orders. This is an advantage to department stores, but it increases their risk of accumulating inventories that may have to be sold at a loss at the end of the season. In order to avoid as much of this as possible and to transfer the risk back to the manufacturer, the stores engage in hand-to-mouth buying, ordering garments in small lots. This is discouraging to apparel firms, and it cuts down on the economies of scale that section work provides. But the mail-order houses, which buy low-priced, more standardized types of garments, are not as concerned with day-to-day changes, and generally place the largest orders, often to be manufactured to their own specifications.

In view of the ease of entry into business and the fact that anyone, including the smallest producer, can come up with a good number that sells, price competition is keen in the industry. The names of only a handful of firms are known to the consumer, who is more concerned with the reputation of the store selling the dress than that of the manufacturer. As a result, the stores often replace the producers' labels with their own. And this, of course, helps to explain the absence of advertising by the apparel houses.

In the high price lines there is some degree of product differentiation among firms, and consumers may be familiar with the names of the top manufacturers. To the degree that this is true, we have a situation of "monopolistic competition," or competition between differentiated products each of which is monopolized by a firm. In these higher price lines, labor costs are not as significant as in the lower-priced merchandise, and therefore firms are not under such great pressure to move out of the established centers of production in search of cheaper labor.

Apparel firms in all price lines try to differentiate their products by selling a garment with greater style-appeal than their competitors' offerings. But since style-pirating is commonplace, they cannot maintain differentiation for very long. They are rapidly faced with price competition, even when they have produced a successful seller. The normal situation, therefore, is that manufacturers and jobbers achieve the tentative beginnings of monopolistic competition, but inexorably, in time, price competition overpowers what they have wrought and they try, try again.

The entrepreneur in the women's and children's apparel industry, dealing with a strong labor union, purchasing cloth from textile firms, and selling garments to retail stores, sometimes sees himself as caught unenviably in the middle. The International Ladies' Garment Workers' Union agrees with this view so far as purchasing and marketing procedures are concerned. The union has asserted that the competitive garment industry is caught between oligopoly (a few big sellers) and oligopsony (a few big buyers).[20]

Though the union's case may be somewhat overdrawn, particularly with regard to oligopoly in the textile industry, we must agree that the apparel manufacturer is indeed surrounded by larger and stronger entities and is vulnerable to their exploitation. The greater the degree of competition and exploitation to which he is subject, the greater the pressure upon him to seek ways of cutting costs. The easiest way, it often seems to him, is relocation.

Before exploring the reasons for relocation, however, one must understand why New York first rose to dominance in this competitive, uncertain, easy-to-enter, low-capitalized, seasonal, fashion-oriented, technologically sluggish industry.

2

New York's Dominance

The women's apparel industry is about 100 years old. Earlier in American history all garments were hand-made and most of them were home-made. Alexander Hamilton, at the outset of the Republic, wrote that in a number of districts two-thirds to four-fifths of all the clothing of the inhabitants was "made by themselves."[1] The well-to-do went to custom tailors and dressmakers, or they imported garments from abroad, which meant England or France.

The production of ready-to-wear started with men's work clothing, either for slaves in the South or for seamen. Brooks Brothers, for instance, started operation about 1830 in New Bedford, Massachusetts, producing ready-to-wear clothes for sailors in that area. As the population grew, others besides sailors, especially bachelors who had no one at home to produce clothing for them, created a demand which was only partly filled by the supply of second-hand clothes; so the second-hand dealers started having clothing made. At first the garments were sewed largely by farm women working at home on material supplied by merchant jobbers. But here and there, workers began to be assembled in groups, and in 1846 the perfection of the sewing machine by Elias Howe opened the way to volume production of ready-made clothing in city surroundings. The Civil War hastened the change. Though the industry lost the plantation business, it nevertheless expanded to meet the big demand for army uniforms; and now the putting-out system rapidly gave way to the factory, using machines and employing mostly women.[2]

Meanwhile the making of women's ready-to-wear was also proceeding—but more slowly. The first report on a women's clothing industry does not appear until the Census of 1860. The products were mainly hoop-skirts and cloaks and mantillas. This industry, once started, grew rapidly. From 1860 to 1880 the value of the products increased from $7 million to $32 million, the number of establishments from 118 to 562, and the number of employees from 5,739 to 25,192.

The women's apparel industry was developed by retail and wholesale dealers. Since they were located primarily in the larger cities—including New York—so was the new industry. The New York dealers, who became the manufacturers, were generally German Jews, and the labor force before 1880 consisted mainly of Irish and German immigrant women, often working at home. In all cities, women dominated the ranks. Although the French tailoresses who cut out the garments were displaced by men after the invention of the cutting knife in 1876, the industry's labor force by 1880 was only 12 per cent male.

The importance of labor

Now came the historic event that ensured New York's dominance in women's apparel. Out of eastern Europe poured hundreds of thousands of Jews. Most were from Russia, including at that time Poland; some were from Austria-Hungary and Rumania. Between 1881 and 1905, Jewish immigrants numbering 850,000 reached the United States. The assassination of Czar Alexander II in 1881, leading to a wave of persecution against the Jews, was one of the immediate causes of this westward movement, but the political motive was only a supplement to deep economic forces that were operating anyhow. As one writer put it, the Jews were "part and parcel of the 'new immigrants' whose appearance on the scene after 1880 was in response to the call of industrial America." [3]

For the immigrant Jews, New York was not only their port of debarkation, but also a center of their kinfolk, where they could live together in a self-imposed ghetto similar to their "shtetels" of eastern Europe. In their communities on Manhattan's East Side and in the Williamsburg and Brownsville sections of Brooklyn, they could converse in Yiddish, enjoy their cultural activities and observe their religious needs without harassment or interference.

The incoming shiploads were arriving in New York harbor at the very time when America was rapidly becoming a highly industrial and urbanized nation, and when the women's apparel industry was expanding phenomenally in an effort to keep up with the demand for its products. By 1890 the industry had 39,149 employees, and by 1900 it had 83,739. The coincidence between the arrival of the immigrants and the upsurge in American consumer demand was the circumstance that permitted New York City to leave all its rivals far behind in the production of apparel.

Just as the merchandising of women's and children's apparel must be located where the buyers congregate, so the location of its production is primarily labor-oriented. That is, the availability of labor is the most important determinant of the location of production. With respect to high-quality, style-changing garments, requiring much hand-work and fancy machine work, this labor orientation is toward skill. In the popular-priced, more standardized clothing lines, where quantity outweighs quality and therefore price competition is intense and the workers need not be as skilled, the labor orientation is toward wages.

The large-scale immigration after 1880 put New York City in a doubly strong position, for it gave the city a skilled labor supply plus a large pool of cheap labor, and the possession of these twin assets was the prime factor in New York's dominance.

Tailoring had been a Jewish occupation in Europe, and therefore some immigrants came as skilled tailors and naturally entered the needle trades. Immigrants who came without skills found that because of the language barrier and their religious restrictions against working on Saturdays, the easiest way to start life in the new country was to go into the trades where many of their coreligionists already were. It was quite common, being considered a moral obligation, for a skilled worker to take his "landsman" from the old country into his shop and teach him the trade. For the less scrupulous skilled worker, "greeners" provided a source of helpers who could be economically exploited during their period of apprenticeship. Eventually those who learned the trade as apprentices came to outnumber their teachers, and since they gained their knowledge after arrival in the United States, they were known as "Columbus tailors."

The entrance of the Jewish immigrants into the clothing industry brought two decided changes.

The first was a great increase in the male proportion of the labor force, since even more Jewish men than women went into the garment shops. By 1890, one-third of the workers in the women's garment industry were men, and in New York the proportion was even higher.

The other change was the spread of the contracting system. Although the beginnings of the system had been visible before 1880 in the form of industrial homework, the manufacturers had kept control over the production, dealing with individual women. This industrial homework gradually evolved into an arrangement in which contractors gathered groups of workers together and acted as intermediaries between the manufacturers and the workers. The contractor was actually an organizer of immigrant labor, and to seek out workers he often met the boats coming into New York harbor.[4] With the advent of contracting, manufacturers kept shops only large enough to handle their normal production—and sent their peak-season production to "outside" shops to be sewed.

Labor conditions had not been good when Irish and German immigrant women in New York City had carried on the work in their homes. They had become worse in the early 1860's with the appearance of tenement houses.[5] And in the 1880's, as more and more workers crowded into the industry, working conditions became more appalling still. The worst conditions were in the mushrooming contractor shops located in the tenement buildings. These were the "sweatshops," in which men and women worked excessively long hours in unsanitary surroundings for extremely low wages.[6]

When the contracting system spread, so did the sweatshop. The contractor's profit was the difference between what he received from the manufacturer and what he had to pay his workers, and he was under inducement to reduce wages as much as possible. If the employer "beat him down," he in turn was forced to "beat down" the wages he paid his workers.[7] Manufacturers played one contractor against another in a kind of auction-block system; the contractors competed among themselves for the prize of survival, and their competition was frightfully hard on the workers.

Tenement and factory laws, beginning in 1892, took work out of the tenements and thus removed some of the worst features of sweating, but the factories were not much better, and it was not until the rise of unionization in the early twentieth century that the sweatshop went into eclipse. At the turn of the century, however, workers continued to labor in dirty, damp, and gloomy places, fifteen to sixteen hours a day, for very low wages. In the 1880's there were sporadic attempts to form unions, but without success.

The growth of the women's clothing industry continued right on through the depression of 1893–1896. In fact, the industry may have even fed upon the national adversity, since the lean times forced many women who had been patrons of custom tailors to turn to the less expensive ready-to-wear.[8] By 1900 the women's clothing industry consisted of 2,701 establishments, turning out $159,000,000 worth of garments. By 1900, too, contracting was declining as the former contractors rose to be small manufacturers. But working conditions remained unsavory.

The continued growth of the industry led to increased demands for labor, and in the 1890's large numbers of Italian immigrants joined the Jewish immigrants in augmenting New York's labor force. By 1900 Italian workers formed about 15 per cent of all the workers in the industry in New York. In Chicago, Cleveland, and other large cities other immigrant groups such as Bohemians, Poles, Russians, and Syrians were drawn into the industry.[9]

Meanwhile the industry was sprouting new branches. For example, the first shop to manufacture shirtwaists was established in

New York City in 1891, and 471 others sprang up there during the next nine years. Unlike cloaks and skirts, this industry used mainly female workers. Another new creation was the women's underwear branch, in which a more detailed division of labor was used and in which women also made up the labor force.

But the single largest part of the industry continued to produce coats and suits, known in the vernacular of the trade as "cloaks." The cloak industry was highly concentrated in New York City where, in 1900, there were 24,000 workers, mostly men, in 830 shops. In that same year, cloakmakers from Manhattan, Brownsville (in Brooklyn), Newark, Philadelphia, and Baltimore held a meeting on Manhattan's East Side. It was no ordinary meeting, for it resulted in a national organization which was chartered by the American Federation of Labor as the International Ladies' Garment Workers' Union. And the ILGWU was to become a powerful force in the functioning of the women's and children's apparel industries.

Garment workers did not immediately flock to the new union, however, and it made little progress until 1909, when it conducted a successful strike in the New York shirtwaist industry, a strike known as the "Uprising of the Twenty Thousand." But the event that firmly established the ILGWU was the cloakmakers' "Great Revolt" of 1910. The union, though small and powerless of itself, called for a strike, and the unorganized New York cloakmakers responded to the call. They stayed out for three months, and finally the intervention of prominent citizens resulted in the employers' recognizing the union. The membership of the New York Cloak Joint Board of the ILGWU skyrocketed to 50,000. Louis D. Brandeis, the future Supreme Court justice, made a basis for the settlement with his "Protocol of Peace," providing machinery for the peaceful solution of labor-management conflicts. Although it disintegrated after six years, the protocol was of lasting significance because it established for the entire women's garment industry the principle of arbitration of labor-management disputes in place of strikes and lockouts.

Both the industry and the union continued their expansion after the cloakmakers' "Great Revolt." Unionization spread to other branches of the industry and to other centers of production, and by 1920 the ILGWU reported a membership of 105,400 out of 165,649 wage earners in the women's clothing industry in the United States. The employers had also organized. To counter the union they had formed in 1910 the Cloak, Suit, and Skirt Manufacturers' Protective Association, and it had become the chief negotiator with the union. As unionization spread, so did the employers' custom of banding together in local industry associations to conduct collective bargaining with the ILGWU.

Though the industry continued to grow, it now did so at a somewhat slower rate, for a simple reason. Until the second decade of the twentieth century its rapid growth had been fed by the continual process of substituting ready-to-wear production for the age-old system of custom tailoring and the sewing of the family's garments by the housewife. But now this process was nearing completion, and the ready-to-wear industry's growth was adjusting to the growth in normal demand for apparel.[10]

During the dynamic period, New York City gathered an increasing share of the industry's production, gaining steadily at the expense of Philadelphia, Chicago, Cleveland, Boston, and other cities. In women's and children's apparel (not including corsets and brassieres and knit outerwear), Census reports showed that New York City's share of the nation's employment reached 62 per cent in 1914, its share of payroll reached 70.5 per cent in 1921, and its share of the value of product reached 74.2 per cent in that same year. Those percentages were about as high as the city ever attained in employment, payroll, and the value of product; but in absolute terms the city continued its vigorous growth.

New York's chief advantage, as we have seen, was its supply of labor, resulting from large-scale Jewish and Italian immigration. But once the immigrants made New York the capital of the industry, the city gained a second precious advantage. It became a giant magnet attracting the buyers of ready-made clothing.[11] Fashion, that strange phenomenon that will lead a woman to discard an otherwise wearable garment simply because it is out of style, became vastly more important in America, beginning in the period during and after World War I. As this occurred, New York became a great fashion center. Its fashion dominance, in turn, made the city "the market" (the place where retailers do their buying from manufacturers and jobbers) for just about every type of women's and children's garment that manufacturers have contrived to produce and consumers to wear.

The role of fashion and the status of Manhattan as a marketplace will be more fully discussed later in the chapter, bringing us up to the present day, but first it is necessary to traverse an eventful and dramatic decade.

The changes of the twenties

Of all the things that happened in the women's and children's apparel industries during the 1920's, perhaps the one with the greatest long-run significance was the shutting off of immigration. Congress imposed restrictions which were aimed precisely at those coming from eastern and southern Europe, the homelands of the clothing workers. It was immigration that enabled New York to replenish and augment its labor force, but now its source of future labor was practically cut off. In other industries—coal and steel, for example—the children of immigrants followed their parents into the mines and mills, but the children of the immigrant garment workers did not do this, except in limited numbers as cutters or pressers, or at entrepreneurial and managerial levels. Manufacturers therefore began to look beyond New York City for new sources of labor.

The full effects of the halting of immigration were not felt right away, however, for three reasons. (1) A large pool of garment workers existed because of immigration up to then. (2) During the depression years of the 1930's there was a slight reversal of the trend for the immigrants' children not to enter the industry. (3) More and more people of Italian extraction, who were either immigrant housewives or their American-born daughters, took the places of the Jewish workers.

A second significant change in the 1920's was the new stress on merchandising. This was related, of course, to the increasing importance of fashion. It was also related to the mounting atmosphere of competition in the industry. During the period in which ready-to-wear production was growing at its fastest, the industry, though competitive, was very profitable. But once the trend toward ready-to-wear production had been completed and supply and demand were in balance, the industry, in order to increase sales, began to concentrate on selling.

The new emphasis on merchandising had an interesting locational result in Manhattan. The industry gradually moved uptown, where it could have fancy showrooms in which to display its merchandise for the buyers. A showroom became more important than salesmen on the road. New York's women's apparel industry had been concentrated on the lower East Side of Manhattan, next to the heaviest colony of Jewish immigrants, which supplied its labor. By the middle 1920's, the Garment Center reached its present—and seemingly permanent—location on the West Side from 34th Street to 40th Street. The northward crawl was abetted by the fact that large numbers of the garment workers, as their status had been raised and they had become better integrated into American life, had moved out of the East Side to settle in the new residential districts of the Bronx and Brooklyn. This dispersal made midtown Manhattan more central to the homes of the workers than the lower East Side, and the new location was better served by rapid transit from all directions.

Another feature of the heavy emphasis on selling was the rise of jobbing and the accompanying decline of "inside" manufacturing. The contracting system in its earlier incarnation had been limited to manufacturers who, when their business exceeded their own shops' capacity, sent batches of cut garments to be sewed in the shops of contractors. The jobber, when he first appeared, however, did not produce at all, but was merely a wholesaler who sent uncut material to contractors to be cut, sewed, and trimmed. With this complete separation of functions, the production process could be carried on at a different location from the merchandising. Furthermore, the rise of the motor vehicle in this decade meant that the location of the production could be fairly distant from that of merchandising.

The jobbers, moreover, had no responsibility for conditions in the contracting shops, and they forced contractors to underbid one another for their orders, pretty much as manufacturers had done in earlier decades. The contractors, as a result, were under pressure to reduce wages.

In the 1920's, therefore, new plants began to spring up in Brooklyn, Westchester County, and northern New Jersey. Back in 1900, there were only 33 women's garment plants in the New York Metropolitan Region outside New York City, and they included 14 plants in the corset industry, which had taken root in New Jersey and Connecticut independently of New York's influence. By 1922, the 33 plants had multiplied to 524. In the same period, New York City's plants rose from 1,823 to 8,455.[12]

The contractors moved in order to escape union control and the higher wages and shorter hours that went with union control. The spread of jobbing weakened the union, already on its knees because of internal warfare; and it was not until the ILGWU's resurgence in the 1930's that the jobber had to assume the responsibility for labor conditions in the shops of his contractors, and that elaborate systems were established for the joint settlement of piece-rate wages. Meanwhile, however, the developments of the 1920's, when contractors began the widening search for communities in which they could outbid other industries for women workers and still pay less than in New York City, were laden with meaning for New York's future.

Although we have now gone far toward setting the stage for the next chapter, in which New York's relative decline in garment production will be described, there remain some further aspects of

New York's national leadership that have not been adequately explored.

The role of fashion

Not only is New York the nation's fashion center, but also its future depends upon the future of fashion as a consumer force. Demand changes in the nation are critical in determining employment in New York's unit-priced dress and coat-and-suit industries, which are the biggest users of labor among the industries covered in this study.

It is one of the most conspicuous facts of American life that the characteristic mode of dress for women is subject to sudden change. A hundred years ago women wore wide skirts reaching to the ground; about 1870 these gave way to tight, body-fitting clothes; then came the era of the bustle and the train; after 1892 the emphasis was on the tailor-made suit. The coming of the bicycle brought shorter skirts. During World War I the skirt became tight and the neckline low; after the war the skirt rose to reveal the knee. During the 1920's the dress replaced the suit as the dominant item of attire, and the dress industry supplanted the coat-and-suit industry as the major branch of women's apparel—a shift, incidentally, that was neutral with respect to location, since both branches were tied to New York. Also in the 1920's, consumer preference switched to silk undergarments, forsaking muslin, batiste, and other cottons, and the negligee became an important apparel item. The flapper's frills eventually faded away and hemlines dropped to more conservative levels, only to rise again. After World War II, Christian Dior's "New Look" again lowered the hem-line, and formal attire began yielding to more casual clothing. Recently the chemise, the trapeze, and the "Empire" have made their appearance. The shape of things to come is beyond an economist's ability to forecast.

A wide diffusion of wealth with increasing prosperity is supposed to accelerate fashion movements. Increasing leisure also affects the sharpness of fashion changes and the rate at which they occur. Such changes were originally limited to the wealthier classes, but today's effective and cheap production of style goods, combined with commercial promotion, makes fashion just about universal.

That New York is the women's clothing capital of the United States is therefore not surprising, even if we ignore the importance of the labor supply. London, Paris, Vienna, and Berlin fill similar roles in their countries. In each case the leader is not only a population center but also the headquarters for elegant society, which sets the tastes in clothing.[13] New York has two other major prerequisites for being a style center: it is both a resort city and an art center.[14] As a resort city it attracts the women who have the daring to wear the new styles, and as an art center it attracts the designers to create them.

New York's artistic qualifications are surpassed only by Paris, and there are those who would place the two cities on a par. New York's museums, particularly of modern art, are outstanding. It has more schools of art and designing than any other metropolis anywhere, and some of these are concerned primarily with fashion design. It is the nation's publication capital, from which emanate the fashion magazines. No wonder that Americans who are interested in any phase of fashion, from textile-fabric design to magazine reporting, gravitate to midtown Manhattan.

Now fashion, of course, is not equally important in all the types of apparel with which this study is concerned. And the differences have a bearing on the geographic distribution of garment-making. The more involved in fashion a garment is, the greater likelihood that it is made in, or near, New York City or one of the smaller fashion centers, for example, Los Angeles or Dallas; and the less fashion-oriented a garment, the farther from the fashion center it can be produced.

Unsurprisingly, a recent market study indicated that consumers, in purchasing children's clothing, place greatest stress on utility and appearance, while in the case of women's apparel, style comes first.[15] Within the women's outerwear industries, the importance of fashion varies from one product to another, and from one price line to another. Unit-priced dresses and coats and suits are highly fashion-oriented, and, for each kind of product, the higher the price of the garment, the more important is styling. Even in the volume-production dozen-priced dress industry, it would be misleading to discount the influence of fashion; a recent meeting of garment manufacturers and designers concluded that the suburban housewives want "chore" or work clothes (house dresses and aprons) with fashion appeal, something more flattering than mere coveralls.[16]

The differences among garments extend also to the speed with which styles change. The more expensive the garment, the greater the possibility that its wearer is a fashion-innovator, or at least is conforming to the standards of a social class which may be regarded as the fashion-setter. Furthermore, even though an expensive garment and a cheap garment may conform equally to the prevailing fashion, the higher-priced one will have a greater degree of individuality, distinguishing it in some way from other similar garments; and it is exclusiveness, as well as fabric and workmanship, which helps to determine price.

Fashion thus moves mainly from high-priced to low-priced lines, though there are exceptions, such as the chemise in 1957 and 1958. Traditionally, as soon as a new style "clicks" at a high price, manufacturers of lower-priced merchandise start making copies of it, using less expensive fabrics and less skilled work. The need to keep abreast of the latest doings in the world of fashion, therefore, requires a firm to locate its headquarters (though not necessarily its production) adjacent to the other producers of fashion goods, and helps to keep such a large proportion of the women's apparel firms concentrated within a few square blocks in Manhattan. A manufacturer producing popular-priced garments for the mass market, though his plants may be distant from Manhattan, may feel an irresistible urge to locate in the Garment Center, where he can either buy designs from model houses or "pirate" them from the style-innovators around the corner. By the time manufacturers in the lowest price ranges are turning out dresses for "mass" consumption, the high-price-line fashion-innovators are exhibiting new styles, which they hope will also gain acceptance.

The economic role of fashion is to promote obsolescence and to induce consumers to purchase new garments beyond their mere replacement needs. From season to season, style changes may seem small, and a woman may be able to "get by" without giving in to a new fashion, but as time passes, her resistance must decrease, lest she become conspicuous by wearing what now is all too manifestly "last year's" fashions.

"The market"

Everything considered, New York no longer occupies the commanding position in the women's and children's apparel industries that it occupied during its World War I peak. But its decline of recent years is a decline in its share of the nation's *production.* Its leadership in *fashion* has continued unimpaired, and its share of the nation's sales has held almost constant since 1939 in some industries and decreased moderately in others.

The dominance of New York as a market place is quite striking. As of January 1, 1956, the area within 30 miles of Manhattan contained the headquarters of 63.7 per cent of the nation's women's and children's apparel firms (manufacturers and jobbers). And these firms' 1955 sales of $3,355,955,000 were 65.7 per cent of the nation's.[17] Actually an even larger proportion of women's and children's garments may be sold from New York, because many firms which have their headquarters elsewhere operate New York showrooms and sales offices, but the sales made in those offices would be credited to the cities where the home offices are.

According to one estimate, about 100 million unit-priced dresses a year are produced for New York-based firms in shops of various types in New York State, Connecticut, New Jersey, Pennsylvania, Massachusetts, Rhode Island, and Delaware, and these dresses constitute 72 per cent of the country's unit-priced dress production.[18] Though the Bureau of the Census no longer lists sales by markets, it does give the volume of shipments by state for 1954. Combining the figures for the states of New York and New Jersey, the great majority of whose shipments emanate from the New York Metropolitan Region, we find that in the women's outerwear industries those two states accounted for the following percentages of the value of national shipments.

Dresses, unit-priced	76.9
Coats	73.9
Suits	66.2
Blouses	64.6
Skirts	56.6
Dresses, dozen-priced (excluding aprons, uniforms, etc.)	29.0

And in some of the intimate lines the percentages were:

Brassieres	60.2
Underwear	53.8
Nightwear	51.1
Corsets	50.4

For knit sweaters the percentage was 60.8. In short, for every item but the highly standardized dozen-priced dresses, the states of New York and New Jersey originated from one-half to three-quarters of the value of the nation's shipments. The two states' shares of shipments are not as large as their shares of sales, because some garments that are sold from New York showrooms are shipped from their point of manufacture.

To see what has been happening to sales in recent years our most convenient measure is the sales figures for New York-based firms —that is, those having their headquarters within a 30-mile radius of Manhattan. The percentage shares of these companies are shown in Table 4 for major groups of garments. Their percentage of the value of national sales of dresses, blouses, and sportswear has remained fairly stationary at about 65 per cent. The coat-and-suit percentage has dropped to 70 from its prewar 75 per cent, but the downward trend seems to have been arrested in 1955. But in the intimate apparel field (underwear, nightwear, corsets, and so forth) a slow but steady decline is evident, with New York firms' sales dropping from 67.9 per cent in 1947 to 62 per cent in 1955. Since style is less important in the intimate garments than in outerwear, the maintenance of a New York headquarters is not as vital. But even if a firm shifts its headquarters from New York to its point

Table 4 Value of Sales of New York-Based Concerns as a Percentage of Total United States Sales for Selected Years

	1939	1947	1948	1950	1954	1955
Women's and children's apparel, total [a]	....	68.8	68.5	67.1	66.1	65.7
Dresses, blouses, sportswear	64.0	66.8 [b]	65.8	64.6	65.7	65.2
Coats and suits	75.0	73.8	74.1	72.5	69.6	70.4
Intimate apparel	....	67.9	67.7	66.8	63.3	62.0

[a] Excludes knit outerwear. The other garments covered by our study are included in the table, though it will be noted that the grouping is somewhat different from that of the Census Bureau.

[b] 1946.

Source: National Credit Office.

of manufacture, the actual sales may continue to be made from Manhattan; the firm either leaves a showroom and sales office behind, or concentrates on production and leaves the merchandising to special sales representatives located in New York.

The drop in the New York-based firms' share of intimate apparel sales is not accounted for by the gains of any other major center,* with the possible exception of Bridgeport-New Haven, which increased its share from 3.8 to 5.0 per cent. Instead, the New York drop is accounted for by the rise of new smaller centers, which have increased their share of national sales from 12.9 to 17.2 per cent.

* The major centers are New York, Chicago, Bridgeport–New Haven, Los Angeles, St. Louis, Allentown–Bethlehem, Philadelphia, and Boston. Bridgeport, it should be remembered, is in Fairfield County, and thus in the New York Metropolitan Region.

In coats and suits, on the other hand, the change in shares has been largely a reshuffling among the major centers, with Los Angeles, Boston, and Kansas City gaining at the expense of New York, Philadelphia, and Cleveland. The post-war trend in consumer demand toward separates and sportswear largely explains the changes, since California has always been a center for casual and sportswear attire, and Boston, a skirt center.

In no case, however, is any other sales center even one-tenth as important as New York, and each exists largely by catering to some specialty to which New York does not. California, by grace of climate, has risen mainly as a center for sport and casual wear; St. Louis is a center for half-size dresses; and the new dress centers of Montreal and Miami specialize in garments to satisfy local seasonal patterns, different from New York's.

Industries servicing apparel manufacturers

The concentration of apparel firms in the Garment Center begets the concentration of other businesses. Many kinds of manufacturing and services ancillary to the women's and children's apparel industries have planted themselves in the same neighborhood. Here are some examples: the design, display, and selling of textiles; sponging (cloth shrinking); factoring (textile banking); trucking; agencies that provide the pretty models; the supplying of thread and trimming; embroidery; the manufacture of belts; and the repairing of machinery.

The presence of these activities in or near the Garment Center provides external economies to the apparel firms and helps to keep them in New York. A firm which has its plant located at some distance from the Garment Center loses these external economies, and it must have a sufficiently large volume of production before it can achieve internal economies. Apparel plants in smaller communities outside New York City, for instance, have to bear the cost of hiring a full-time machinist to fix sewing machines when they break down, even though they cannot keep him fully occupied. A firm with a shop in the Garment Center, on the other hand, can call a repair man from down the street and pay him only for the time actually spent in repairing its machines. The distant plants also often experience delays in production, while they await deliveries of needed sewing machine parts, whereas the Garment Center shop can order spare parts from around the corner for immediate delivery.

The external economies of the Garment Center explain why garments sewed outside the city are returned to it for final processing. Only when the volume of business becomes large enough to justify doing one's own embroidery or manufacturing one's own belts, on the premises, can integrated operations take place. Since so many apparel firms are small, they cannot achieve these internal economies, and they remain dependent upon the external economies of the Garment Center. Manufacturing industries that are heavily concentrated within the New York Standard Metropolitan Area because of their close connection with apparel production—in some cases, men's as well as women's and children's—are shown in Table 5.

Table 5 Manufacturing Industries Accessory to Apparel, New York Standard Metropolitan Area, 1954

Industry	Employment in area	Percentage of U.S. employment
Belts	5,000–9,999 [a]	[a]
Pleating, stitching, and tucking for the trade	4,593	71.2
Schiffli-machine embroideries	4,934	86.3
Embroideries, except Schiffli-machine	8,613	79.7
Sewing machines	6,828	55.0
Buttons	4,152	49.0
Needles, pins, hooks and eyes, and similar notions	7,272	31.4

[a] Precise data for the belt industry in the Standard Metropolitan Area are not available for 1954. In 1947, however, the metropolitan area accounted for 53.8 per cent of the employment of the belt industry in the United States. In 1954, the belt industry in the United States had 12,571 employees.

Source: U.S. *1954 Census of Manufactures.*

Just where, within the New York Metropolitan Region, the ancillary services locate depends upon their need for direct face-to-face communication. Those services requiring constant and direct contact between servicer and apparel firm are tied to Manhattan's Garment Center, while the others are able to locate elsewhere in the city or Region. In order to achieve closer direct contact, New York's textile district is now in the process of physically relocating from downtown Manhattan to the heart of the Garment Center. Of the 34 textile designers listed in a New York industrial directory, 23 have addresses in Manhattan postal zones 1 and 18, which run from 25th to 41st Streets west of Fifth Avenue, thus approximating the Garment Center. The same is true of 119 of the 170 jobbers of woolens, not including those dealing exclusively with men's wear fabrics; 104 of the 118 firms selling neckwear trim and belts; 64 of the 110 sellers of sewing threads; 51 of the 69 firms selling snap fasteners, adjustment devices, weights, pins, hooks and eyes, and such items.[19]

Embroidery, tucking, pleating, and stitching are all trim to the finished garment, and hence are located within or adjacent to the Garment Center. Most of belt production is in a similar situation; the Beltmakers' Union reports that 3,500 of its 5,100 members are employed making belts for dresses, skirts, and other garments.[20] The Schiffli-machine embroidery industry, however, is concentrated at Union City, in Hudson County, New Jersey, across the Hudson River from the Garment Center.[21] Most of the Region's employment in the sewing machine industry is accounted for by the Singer Sewing Machine Company's plant in Elizabeth, New Jersey. Singer also has another large plant in Bridgeport (Fairfield County, Connecticut) at which it produces exclusively industrial sewing machines. Although Singer's location in the Region preceded even New York's dominance in the apparel field, its continued presence is undoubtedly influenced by the heavy concentration of garment manufacturing close by.

In this chapter we saw how New York City rose to dominance as a production center of women's and children's apparel because immigration brought it a large pool of cheap labor and a skilled labor supply at the very time that the industry was emerging on a large scale. The 1920's saw the beginning of forces that were to weaken New York's rule as a production center. By then, however, the city had become the nation's great design and sales center for women's apparel, and the important role of fashion has permitted it to retain that position.

Now we must turn to the question of the production decline, and how and why it took place.

3

New York's Decline as a Production Center

The island of Manhattan is still the source of more women's and children's apparel than any other locality on earth. It still leads not only in the design and the merchandising but also in the production—the sewing of garments. But the production has been noticeably outward bound, in the sense that areas outside Manhattan have experienced faster growth. Within New York City there has been diffusion from Manhattan to Brooklyn; within the New York Metropolitan Region, from the city to the outlying counties; and on a broader canvas, from the Metropolitan Region to Pennsylvania, New England, and the South.

As a result of the production shifts, the Region's share of national employment in the women's and children's apparel industry has long been declining, and now the Region seems to have passed the peak of its employment in absolute terms as well. The forces that pulled or drove production outward from the Garment Center will be examined in detail in this chapter.

THE EMPLOYMENT TRENDS

✓ The Region's relative losses

A few paragraphs will suffice to give some idea of the extent of the Region's decline, the industries that were principally involved, and the areas that gained at New York's expense.

In 1914, when the Region's garment shops were concentrated

even more heavily in Manhattan than they are today, New York City could claim 62 per cent of the nation's wage earners in the "women's clothing industry," a classification roughly comparable to the industries covered by this report. By 1939 the figure had dropped to 41 per cent.[1] What had happened was that in the 1920's and 1930's the more standardized products, less concerned with style, began to take root elsewhere, especially in the anthracite area of Pennsylvania with its large supply of female labor. During those decades, however, the New York Metropolitan Region held tightly to the production of women's outerwear, particularly the two main branches of women's outerwear, one being unit-priced dresses and the other being coats, suits, and skirts. These two industries are not only the largest but also the most style-oriented of the twelve industries we are studying.

World War II temporarily arrested the erosion of the Region's share of the apparel industries, because consumer demand increased greatly and manufacturers were willing to produce wherever they could find facilities and workers. In 1947, New York City still accounted for about 40 per cent of the national employment. And the New York Standard Metropolitan Area of 17 counties accounted for 48.7 per cent.[2]

By the middle 1950's, however, a considerable drop in these percentages had occurred, as shown by all available evidence. In fact, a glance at the last two columns of Table 6 indicates that the share of the metropolitan area decreased in each of the twelve industries with one exception, the corset-and-brassiere industry.[3]

Exactly what happened to the Region's employment in absolute numbers after 1947 is blurred by somewhat conflicting statistics. The federal manufacturing census of 1954, which is the basis for Table 6, suggests that in absolute numbers the Standard Metropolitan Area held its own, and even increased its employment, in our twelve industries as a group, though not in each of them.

But a compilation for 1956, based primarily on state employment statistics, indicates that by this time a decline had appeared not only in the area's share but also in the absolute level of its employ-

Table 6 Employment in the Women's and Children's Apparel Industries, United States and New York Standard Metropolitan Area, 1947 and 1954

	Thousands of employees				Ratio, New York area to U.S.[b]	
	United States		New York area[a]			
	1947	1954	1947	1954	1947	1954
Total, all industries	490.6	600.1	239.0	260.8	48.7%	43.5%
Women's and misses' outerwear, total	313.1	363.9	164.0	168.7	52.4	46.4
Blouses	32.9	43.0	16.7	17.3	50.7	40.2
Dresses, unit-priced	133.8	143.3	82.6	81.9	61.8	57.2
Dresses, dozen-priced	48.6	54.5	6.3	6.8	13.0	12.5
Coats, suits, skirts[c]	82.7	96.0	51.2	53.4	61.9	55.6
Neckwear	2.8	1.7	2.5	1.5	89.7	84.7
Other outerwear	12.4	25.4	4.6	7.8	37.2	30.6
Women's and children's undergarments, total	94.6	112.2	37.8	39.2	39.9	34.9
Underwear, nightwear	58.0	73.4	24.4	23.5	42.1	32.1
Corsets, brassieres	36.6	38.8	14.3	15.7	39.0	40.4
Children's outerwear, total	48.2	77.5	22.7	33.7	47.1	43.5
Dresses	22.5	32.5	8.9	12.4	39.8	38.2
Coats	11.0	14.4	7.8[d]	8.6	70.6	59.7
Other outerwear	14.8	30.6	6.3[d]	12.7	42.9	41.5
Knit outerwear	34.5	46.4	14.6	19.2	42.2	41.4

[a] Constituent industries may not add exactly to wider industry classification totals because 1947 figures for women's outerwear, women's and children's underwear, and children's outerwear were revised, but there is no precise way of reallocating the revisions by constituent industries.

[b] Percentages are based on actual figures rather than those presented here which are rounded to the nearest hundred workers.

[c] An interesting result is obtained by separating the coat-and-suit employment from the skirt employment. In coats and suits the U.S. employment declined from 74.4 thousand to 74.2 thousand; the New York area's employment declined from 45.7 thousand to 41.2 thousand; and the area's share declined from 61.4 to 55.6 per cent. On the other hand, U.S. skirt employment rose from 8.4 thousand to 21.8 thousand; the New York area's skirt employment rose from 5.6 thousand to 12.1 thousand; and the area's share declined from 66.6 to 55.6 per cent. Since the Census Bureau did not give separate figures for 1954, our 1954 breakdown on a U.S. basis was obtained by applying manufacturer and jobber breakdowns to the totals. The breakdown for the New York area was obtained by using the same ratios as in the U.S. breakdown.

[d] 1947 figures revised because children's skirts were transferred from coat industry to other outerwear industry in 1954. Ratio of New York State and New Jersey revised figures applied to New York Standard Metropolitan Area.

Source: U.S. *Census of Manufactures*, 1947 and 1954.

ment in these industries. Table 7 gives a comparison between certain months of 1947 and the same months of 1956 in the New York Metropolitan Region, which, as pointed out earlier, includes five more counties than the metropolitan area covered by the federal statistics. According to this table the Region by 1956 had suffered absolute loss of jobs in almost all the components of women's outerwear. It had also lost jobs in the underwear-nightwear industry. To some extent these losses were offset by gains in corsets and brassieres and in all branches of children's outerwear. The Region made a really impressive gain in the knit outerwear industry, so that the net loss of jobs in all twelve industries was only 1,347. If one were to exclude knit outerwear (which, as we have noted, the government classifies as a part of the textile industry) the Region would be left with a net loss of 11,638 jobs in the other eleven industries.

One of the most striking developments of the post-war period is that the two big style-oriented industries, unit-priced dresses and coats, suits, and skirts, joined other apparel industries in the parade away from the Region. According to Table 7, the Region's greatest job losses, by far, were in those two industries; between 1947 and 1956 their combined job strength in the Region fell from 155,024 to 136,167.*

The New York area's losses in style products were principally in the cheaper ends of those products, and somewhat in the medium price ranges. Payroll statistics for unionized shops illustrate this point readily in the case of unit-priced dresses. Table 8, based on union sources, shows that work performed for New York City firms tended to stay tenaciously in New York City if the dresses

* The federal figures in Table 6, collected under different procedures and computed for a different time period and a different number of metropolitan counties, show an increase of 1,500 jobs in the two style industries—almost stationary—from 1947 to 1954. Moreover, the area's employment in these two important industries would have shown a marked decline even according to the federal statistics, if it had not been for a gain of 6,500 jobs in a single garment—skirts—a garment which is rapidly increasing on a national basis. (See Table 6, footnote c.)

Table 7 Average July-and-September Employment in Women's and Children's Apparel Industries, New York Metropolitan Region, 1947 and 1956[a]

	1947	1956	Change 1947–1956
Total, all industries	261,150	259,803	−1,347
Total, excluding knit outerwear	247,155	235,517	−11,638
Women's and misses' outerwear, total	182,572	163,325	−19,247
Blouses	15,765	14,182	−1,583
Dresses, unit-priced	90,213	78,765	−11,448
Dresses, dozen-prized	3,718	3,684	−34
Coats, suits, skirts	64,811	57,402	−7,409
Neckwear	2,751	2,365	−386
Other outerwear	5,314	6,927	+1,613
Women's and children's undergarments, total	40,790	41,682	+892
Underwear, nightwear	25,204	24,450	−754
Corsets, brassieres	15,586	17,232	+1,646
Children's outerwear, total	23,793	30,510	+6,717
Dresses	9,148	10,151	+1,003
Coats	7,348	8,723	+1,375
Other outerwear	7,297	11,636	+4,339
Knit outerwear	13,995	24,286	+10,291

[a] Adequate figures available for July and September only. This table compares the average of those two months in 1956 with the average of the same two in 1947, except that for Fairfield County, employment in both years is for first quarter. Employment in some industries, for instance, coats, suits, and skirts, is seasonally high in July and September, and this fact helps to explain variations from U.S. Census employment figures, which are 12-month averages. Employees included in this table are those covered by state unemployment insurance programs—slightly less than total employment.

Source: Unpublished data from the departments of labor of New York State, New Jersey, and Connecticut, except that the 1947 figures for Dutchess, Putnam, and Orange Counties are our own estimates and the 1947 figure for Fairfield County is estimated from U.S. Census Bureau, *County Business Patterns, First Quarter 1947*. The figures for all New York counties had to be adjusted slightly for comparability; for example, the basis of collecting data changed between 1947 and 1956.

were expensive, but tended to move out of town fast if the dresses were only a few dollars apiece. And "out of town," in this industry, means principally outside the New York Metropolitan Region.

Most of the relative decline suffered by the Region in the women's outerwear industries has been taken up by areas within over-

Table 8 Percentage Distribution of Production-Worker Payroll in New York Production Area,[a] Unit-Priced Dress Industry, by Dress Price Lines, 1946 and 1956

Dresses by wholesale price line	Percentage in N.Y. City shops		Percentage in out-of-town shops	
	1946	1956	1946	1956
All prices	84.57	71.49	15.43	28.51
Under \$5.75	78.14	40.21	21.86	59.79
\$5.75–\$16.75	81.84	62.86	18.16	37.14
Over \$16.75	98.68	94.30	1.32	5.70

[a] "New York Production Area" includes unionized manufacturers, jobbers, and contractors in New York City plus all out-of-town dress establishments performing work for them. Most of the out-of-town shops in this industry are outside the New York Metropolitan Region—for example in Pennsylvania and Massachusetts. In out-of-town shops which perform only part of their work for New York City's unit-priced dress firms, only that part of their payroll is covered by this table.

Source: New York Dress Joint Board, ILGWU.

night trucking distance of the New York style center. In fact, in the outerwear industries it is useful to think of the five states of New York, New Jersey, Connecticut, Massachusetts, and Pennsylvania as a group, which, for convenience, can be called the "New York Production Area." This group has just about maintained its share of the nation's employment in the women's outerwear industries in the post-war period: 71.8 per cent in 1947 and 70.6 per cent in 1954.[4] This stability is clearly illustrated by the biggest outerwear industry, unit-priced dresses, in which these five states in 1939 had 79.9 per cent of national employment; in 1947, 80.1 per cent; and in 1954, 80.6 per cent. But within the production area there has been anything but stability. The center of the area, New York City itself, has been losing to the fringes; the new plants sewing garments for New York manufacturers and jobbers have been established mainly in Pennsylvania, and, to some degree, in Massachusetts. In some style products, there has been a rise of new fashion centers since the 1930's—such as those in California, Missouri, and Texas in coats, suits, and skirts. Even in this industry, however, the five-state New York Production Area maintained its share of national employment between 1947 and 1954.

In undergarments and children's outerwear, which are less style-oriented than women's outerwear, production has been able to move beyond overnight trucking distance of Manhattan; and not only the New York Metropolitan Region but also Pennsylvania has declined in its share of national employment. These relative losses have been the South's gain. Between 1947 and 1954 the Southern states increased their share of national employment in undergarments from 8.4 per cent to 19.3 per cent, and in children's outerwear from 7.9 per cent to 14.1 per cent.[5] In fact the South has gained from other areas of the nation as well as from the northeastern states. It has also enjoyed a moderate increase in the production of women's outerwear, particularly of the more standardized products, including dozen-priced dresses (household apparel) and sportswear items, such as slacks.

Trends within the region

With few exceptions, the apparel plants in the outlying counties of the New York Metropolitan Region—that is, outside New York City—either are owned by Manhattan manufacturers or are the shops of contractors working for Manhattan jobbers. As production has moved outward from New York City, these outlying counties of the Region have increased their employment, though not quite enough to compensate for the city's losses. Statistics from the governments of the three states which the Region straddles indicate that between 1947 and 1956, New York City's employment in our twelve industries decreased by 12,201 jobs while the employment in the rest of the Region was increasing by 10,854. Table 9 shows the employment changes in each industry for the city and for the rest of the Region. During this same nine-year period, New York

Table 9 Average July-and-September Employment in Women's and Children's Apparel Industries in New York City and Rest of New York Metropolitan Region, 1947 and 1956

	New York City			Rest of Region		
	1947	1956	Change 1947–1956	1947	1956	Change 1947–1956
Total, all industries ...	208,398	196,197	−12,201	52,752	63,606	+10,854
Women's and misses' outerwear, total ..	149,213	127,204	−22,009	33,359	36,121	+2,762
Blouses	12,982	11,519	−1,463	2,783	2,663	−120
Dresses, unit-priced .	75,511	64,420	−11,091	14,702	14,345	−357
Dresses, dozen-priced	2,532	2,608	+76	1,186	1,076	−110
Coats, suits, skirts ..	52,252	41,578	−10,674	12,559	15,824	+3,265
Neckwear	2,393	2,279	−114	358	86	−272
Other outerwear ...	3,543	4,800	+1,257	1,771	2,127	+356
Women's and children's undergarments, total	30,036	29,514	−522	10,754	12,168	+1,414
Underwear, nightwear	19,540	19,387	−153	5,664	5,063	−601
Corsets, brassieres ..	10,496	10,127	−369	5,090	7,105	+2,015
Children's outerwear, total	18,307	21,220	+2,913	5,486	9,290	+3,804
Dresses	6,618	6,426	−192	2,530	3,725	+1,195
Coats	5,669	5,874	+205	1,679	2,849	+1,170
Other outerwear ...	6,020	8,920	+2,900	1,277	2,716	+1,439
Knit outerwear	10,842	18,259	+7,417	3,153	6,027	+2,874

Note: Totals for Region are in Table 7. For explanation of months used, employees covered, and sources, see notes to Table 7.

City's share of the Region's employment in the twelve industries dropped from 80 to 76 per cent.

We have seen that the outward movement in unit-priced dresses occurred mainly in cheaper and medium-priced merchandise and that the gains were made in areas outside the Region. In coats and suits, which are heavier garments and require a greater degree of skilled hand labor, the outward movement has also been mainly in the cheaper ends, but the production has not moved so far. More of it has stayed within the Region. The outlying counties gained 3,265 jobs in the coat-suit-and-skirt industry during this period.

The distribution of employment between the city and the rest of the Region varies by industry, neckwear being the most heavily concentrated in the city and corsets and brassieres the least. Although most of the apparel growth in the Region outside the city represents migration from Manhattan, some of this manufacturing has always been there. The heavy concentration of corset-and-brassiere production in the outlying counties is as old as the industry itself, for Bridgeport and northern New Jersey have always been corset centers.

Table 10 gives an idea of what counties of the Region outside New York City have the largest employment in the women's and children's apparel industries, and what counties are growing the fastest.

Hudson County, New Jersey, has by far the most apparel jobs among these counties. Moreover, it was still growing in this respect between 1947 and 1956, though it was no longer a place of population growth. Hudson County, like Brooklyn, is just across a river from Manhattan, and thus offers a suitable location for manufacturers who do not want to—or cannot—maintain their production and distribution in Manhattan, yet wish to keep it close to their showrooms.

Suffolk County, occupying the eastern expanses of Long Island, was the leader in rate of growth, nearly tripling its employment between 1947 and 1956. Its next-door neighbor, Nassau, also expanded fast. Long Island's apparel growth was due largely to population growth, since the industry is continually seeking new sources of female labor. Proximity to Manhattan, enhanced by construction of the Long Island Expressway, should lead to further expansion of apparel employment on the island, particularly in Suffolk.

Significant gains were also recorded in the counties on the northern tier of the Region—Dutchess, Putnam, and Orange. The industry in that tier is clustered around Newburgh in Orange County, and, to a lesser extent, around Poughkeepsie in Dutchess County. The New York State Thruway, and other transportation improvements to come, should make the area even more attractive for apparel growth in the future.

Table 10 Employment Growth in Women's and Children's Apparel Industries in the Counties of New York Metropolitan Region Outside New York City, 1947 and 1956[a]

	Employment		Absolute change	Percentage change
	1947	1956		
All counties outside city	52,572	63,606	+10,854	+20.6
New York State counties	11,002	14,330	+3,328	+30.2
Nassau	1,417	2,149	+732	+51.7
Suffolk	932	2,741	+1,809	+194.1
Westchester-Rockland	5,673	5,368	−305	−5.4
Dutchess-Putnam-Orange	2,980	4,072	+1,092	+36.6
New Jersey counties	37,721	44,371	+6,650	+17.6
Hudson	10,526	13,684	+3,158	+30.0
Essex	7,980	7,321	−659	−8.3
Passaic	5,405	7,512	+2,107	+39.0
Bergen	3,617	4,299	+682	+18.8
Union-Morris-Somerset	3,153	3,924	+771	+24.5
Middlesex	4,152	4,764	+612	+14.7
Monmouth	2,888	2,867	−21	−0.7
Connecticut (Fairfield County)	4,029	4,905	+876	+21.7

[a] Average of July and September. For explanation and sources, see Table 7.

On the other hand, the Westchester-Rockland area, which lies between the northern tier and New York City, has actually declined a bit in apparel jobs. Westchester County became an important dress production center in the 1920's, when the industry began to leave Manhattan in search of less expensive female labor. In time, unionization caught up with employers who had moved to Mount Vernon and other places in Westchester, and the reduction of differentials in wages caused the area to lose one of its major attractions for apparel manufacturers. Too, available job opportunities in other industries have reduced the potential labor force for the apparel industry. Except for New York City the Westchester-Rockland area is still the most important part of the Region in the unit-priced dress industry, but competition from other industries with rising wage rates makes it unlikely that Westchester and Rockland will increase their apparel employment in the future.

Essex County, New Jersey, whose chief city, Newark, has historically been a small apparel center in its own right, is another declining county. But Passaic County to the north registered a healthy post-war increase, as did its neighboring counties in northern New Jersey. The Jersey counties, including Hudson and Essex, are especially strong in coats, suits, and skirts.

The reader who wants to know how many jobs each county had in 1956 in various major apparel industries (such as unit-priced dresses and women's and children's undergarments) will find a suitable table in Appendix B. Meanwhile our present business is to begin examining more closely the reasons for the production decline of the New York Metropolitan Region.

SHIFTS IN CONSUMER DEMAND

The Region's share of national employment could change either because of a relative change in demand for the things that it produces best or because of relative changes in its cost position. Both kinds of changes have in fact occurred. First we shall examine the effects of shifts in consumer demand, and then take up the crucial question of labor cost.

The changing pattern of American life that has emerged in the last decade has brought about a "revolution in clothes."[6] The tremendous growth of suburban communities has led to a more casual way of living, with informal dress—such as separates, dungarees, shorts, playsuits, slacks, and other sport-type garments—replacing dresses and coats and suits.

The shift to casual wear is illustrated in Table 11, showing large post-war increases in the production of skirts, blouses, and dozen-priced (household) dresses, much smaller increases in coats and unit-priced (streetwear) dresses, and an actual decline in women's suits.

The New York Metropolitan Region was adversely affected by the shift in the types of garments wanted by the consumer, because it has been a more important production center of formal types of

Table 11 Changes in Production of Selected Items of Women's and Misses' Outerwear, United States, 1946 to 1955

	Thousands of units produced		Percentage change
	1946	1955	
Informal clothes:			
Skirts	21,540	82,812	+284.5
Blouses	73,848	169,596	+129.7
Dozen-priced dresses	78,888	126,180	+59.9
"Formal" clothes:			
Suits	17,491	15,651	−10.5
Coats	21,824	25,607	+17.3
Unit-priced dresses	134,185	146,002	+8.8

Source: U.S. Bureau of the Census, *Facts for Industry* series for women's outerwear.

apparel than of the casual, sportswear types. In 1954 the states of New York and New Jersey (overwhelmingly accounted for by the New York Metropolitan Region) shipped 76.9 per cent of the nation's dollar value of unit-priced dresses, but only 29.0 per cent of dozen-priced dresses. In coats the percentage was 73.9 and in suits 66.2, but in skirts it was only 55.6. The two states shipped 64.6 per cent of the nation's blouses, by dollar value, but much of the *production* of this casual item is carried on in the shops of Pennsylvania contractors working for New York jobbers.

Had there been no shift in the "mix" of products purchased by consumers, the New York Metropolitan Region, we estimate, would have had about 3,000 more workers in women's and misses' outerwear in 1954 than it actually had. Even this does not tell the entire story, because within each industry of women's outerwear, some products are more casual than others, but it is impossible to get accurate data within the industry. Nevertheless, it is clear that the New York apparel industry has suffered relatively from the trend toward casual garments.

Informal living and the "do it yourself" movement have also resulted in a demand for cheaper, more utilitarian attire. This demand, like the demand for sportswear, has adversely affected New York producers, which concentrate on the higher-priced garments. Between 1948 and 1953 the average wholesale value of coats and suits decreased 15.6 per cent, and of unit-priced dresses 5.3 per cent.[7]

The post-war period has also seen the rise of new synthetics, such as nylon, orlon, and dacron. In some cases, where the switch has been from expensive to cheaper fabrics, this development has hurt New York's production. New York manufacturers have not been as free as others to switch to less expensive fabrics because such action would require them to lower the price of their products in order to remain competitive. Once in a lower-priced field, however, they could not match the wage costs prevailing in competing areas outside New York, as we will see later.

New York coat-and-suit manufacturers, for instance, generally use wool in their garments, but part of the post-war trend toward casual wear was the substitution of lighter-weight man-made fibers for wool in the production of suits, coats, and snow suits. Although wool has made some comeback since 1952, it still trails far behind its 1946 usage.

Shifts in fabrics have hurt New York production in another way: they have enabled producers located elsewhere to produce garments formerly in the price domain of producers concentrated around New York. In dresses and blouses the most important fabric shift has been from rayon to cotton. This trend has been facilitated by several technological developments in cottons, including wrinkle-resistant finishes, permanent glazes, embossed effects, and fabrics requiring no ironing. By 1955 six out of ten blouses were made of cotton, as against less than one out of ten in 1946. By 1955 three out of ten unit-priced dresses were being made of cotton, whereas in the 1930's nine out of ten had been made of rayons and acetates. Dozen-priced dresses have always been dominated by cotton, and in 1955 almost nine out of ten were made of that material. Because of the improvements in cotton fabrics, producers in the South and elsewhere have been able to turn out more attractive and expensive garments with their relatively unskilled labor, thus overlapping the price lines of producers in the New York Metropolitan Region.[8]

Thus the shifts in fabrics have brought increased competition as between one industry and another—as, for instance, between dozen-priced dresses and unit-priced dresses, or between coats-suits-skirts and women's "other outerwear." This competition is aggravated by the fact that the different industries are organized by different locals of the ILGWU, each operating under a collective agreement specifying certain wage scales. As a result some industries have lost production to others whose wage scales are lower than theirs; and this trend has induced firms in the higher-wage industries to try to restore their competitive position by shifting production out of New York to lower-wage areas.

The very simplicity of the garments which have risen in the post-war period has an important bearing on New York production. One of New York's greatest advantages as a production center has been its supply of skilled labor, able to turn out the fanciest sewing by machine and hand. As apparel is simplified there is less need for hand-work, and New York's advantage declines. When garment-making consists of little more than joining a few seams together, production can be carried on at a wide choice of locations, because the labor force for simple tasks can be quickly trained and the major attraction of any location becomes its supply of cheap female labor. Thus the discussion of consumer demand has led us back to the major question of labor.

THE SEARCH FOR CHEAP LABOR

The relative decline of the New York Metropolitan Region as a center for the production of women's and children's apparel began long before the post-war changes in consumer tastes. The decline has been due primarily to a cost disadvantage, and most of all a labor-cost disadvantage, in the manufacture of less expensive, more standardized garments. The supply and cost of labor are the most important factors influencing the location of the industry. In order to understand the role that labor has played, we need to examine not only wage differentials but also certain other developments.

The use of less skilled workers

The outward movement of apparel production has been made possible by two significant developments. First, the separation of sewing from merchandising enabled the garment entrepreneur to enjoy the best of two possible worlds. He could locate his showroom in Manhattan's Garment Center and his production outside the city wherever production costs were lowest. Second, the advent of the elaborate divisions of labor known as section work ended the dependence of production upon New York's supply of skilled labor. In section work, the employer uses a core of skilled workers for only the more intricate operations while the bulk of the labor force performs routine and simple sewing.

Section work was introduced early in this century in those branches that were less susceptible to the dominance of style, such as underwear. This was true both in New York and elsewhere. But section work was not so readily applicable to the style products—unit-priced dresses and coats and suits—because few of any particular pattern were produced. By the 1920's, however, section work made its appearance in the style industries, too, but only outside New York, and in the cheaper lines where the volume of orders made it economically feasible.

Section work forged ahead steadily during the 1920's and 1930's. It gained even more during World War II, when, in the absence of other outlets for consumer income, there was a great increase in demand for women's apparel, cutting across all price lines. New York City, with its large supply of skilled tailor-system machine operators and hand sewers, concentrated on what it was best at—

the manufacture of medium-priced and higher-priced merchandise—and producers of less expensive garments sought facilities and workers elsewhere. In the case of the style products, where speed is essential, it was necessary to keep production within overnight trucking distance of Manhattan showrooms, and new establishments cropped up at various points in New Jersey, upstate New York, Pennsylvania, and Massachusetts.

Within a few years after the war ended, demand for apparel slackened considerably and the industry once more became highly competitive. And the new plants that had sprung up outside New York City, using less skilled, lower-paid workers, enjoyed a competitive advantage in the production of lower-priced merchandise.

Section work had not grown very much in the style industries within the city, for numerous reasons. The tailor system was more adaptable to the higher-priced lines in which New York City specialized. But even in the lower-priced lines there was little growth of section work in the city. Section work requires more workers per shop and therefore more space per shop, and large plant space was not readily available in Manhattan, particularly in the Garment Center. (Some large space was available in other boroughs, and section work plants did arise in Brooklyn.)

Another reason for the lag of section work in the style industries in New York City was the opposition of the workers to this system. The older workers considered themselves in the skilled category, and were reluctant to permit a diminution of skill.[9] Since they controlled the policies of the union, they were able to cause it to use its power to prevent the large-scale introduction of section work in the Garment Center. The general manager of the Cloak Joint Board has observed:

> It is fruitless to discuss whether or not any steps could have been taken to direct the growth of section work in the proper places and under different conditions. For understandable reasons, the union, in the past, took a position against section work, even refusing officially to recognize it, even though it existed right across the bridges over the East and Hudson Rivers.[10]

Today section work is dominant in most branches of apparel, and only the dress and coat-and-suit branches have not been engulfed by it. And in those branches, it is only the older centers of production that still use the tailor system to a large degree.

THE PRICE OF LABOR

As the women's and children's apparel industries adapted themselves to increasing use of unskilled labor, wage differentials among geographical areas assumed more and more importance. Labor is a substantial part of total cost in the garment industry. It is not so costly as fabric, which is the biggest item of all (see Table 12), but

Table 12 Percentage Distribution of Costs in Women's and Children's Apparel Industries, United States

Industry	Total costs	Wages of production workers	Material	Manufacturers' overhead and profit
Dresses	100	27.0	46.0	27.0
Blouses	100	27.4	45.2	27.4
Coats, suits, skirts	100	23.8	50.8	25.4
Children's outerwear ..	100	23.8	50.8	25.4
Corsets, brassieres	100	20.0	43.6	36.4

Source: ILGWU Research Department in *Trends and Prospects, Women's Garment Industry, 1953–1956.*

the cost of fabric is the same for all firms, except very large purchasers. Since establishing a shop outside New York City does not alter the firm's fabric cost, the most important variable cost in terms of inter-regional competition is labor.

New York City is a high-wage area for the women's garment industry. This fact is reflected in the statistics on average earnings given in Table 13, which shows that the city's production workers average weekly and hourly earnings significantly higher than the national averages. New York's higher levels hold true not only for the aggregate of all apparel employees but also for each branch of apparel—for example, $2.54 an hour in coats, suits, and skirts as

Table 13 Average Weekly and Hourly Earnings of Production Workers in Women's and Children's Apparel Industries, United States and New York City, 1956

	United States		New York City		U.S. earnings as percentage of N.Y. City earnings	
	Weekly	Hourly	Weekly	Hourly	Weekly	Hourly
Women's outerwear	$55.42	$1.57	$73.05	$2.14	75.87%	73.36%
Dresses, unit-priced	55.62	1.58	71.87	2.10	77.39	75.24
Dresses, dozen-priced	44.76	1.24	....	....	....	....
Coats, suits, skirts	67.94	2.01	87.13	2.54	77.98	79.13
Women's and children's undergarments	47.92	1.32	54.20	1.56	88.41	84.62
Underwear, nightwear	45.38	1.25	54.42	1.54	83.39	81.17
Corsets, brassieres	51.77	1.43	53.75	1.60	96.32	89.38
Children's outerwear	48.44	1.32	56.31	1.65	86.02	80.00
All above industries	52.83	1.48	67.69	1.98	78.05	74.75
Knit outerwear [a]	56.30	1.47	65.77	1.73	85.60	84.97

[a] Knit outerwear could not be included in total for United States because data on the number of employees in that industry were not available. New York City figures for all industries in table, including all knitting mills (these are overwhelmingly knit outerwear mills), would be $67.52 and $1.95 for weekly and hourly earnings, respectively.

Sources: Based on data of the U.S. and New York State Departments of Labor.

compared with the national average of $2.01, and $1.56 an hour in undergarments as compared with $1.32. To be sure, there has been a lessening of differentials in earnings between New York and other major cities, but it is not to these other metropolitan centers that New York has been losing production; it is to small cities and towns. New York City's wage position in the mid-1950's remained almost as high, when compared with that of the nation as a whole, as it had been shortly after the war; in the few women's garment industries for which reliable comparisons could be made, earnings in New York City exceeded those of the United States by roughly the same percentages in 1955 as they did in 1947.*

* Based on average hourly earnings provided by the U.S. Bureau of Labor Statistics for unit-priced dresses, underwear-nightwear, and knit outerwear.

One of the most meaningful comparisons that can be made is between New York City and the anthracite district of Pennsylvania in the unit-priced dress industry. In August 1955, according to the federal Bureau of Labor Statistics, production workers in that industry had average hourly earnings of $2.16 in New York City as against only $1.13 in Luzerne County, Pennsylvania, which includes the many dress plants of Wilkes-Barre and Hazleton.[11] The New York figure, however, includes earnings in both regular and contract shops in all price lines, whereas all the Pennsylvania shops studied were contract shops and the area is one of inexpensive merchandise. In those New York City dress factories that are most comparable to the ones in Luzerne County—that is, contract shops producing dresses that wholesale for less than $6.75—the average hourly earnings were $1.68. The Pennsylvania figure of $1.13 was thus 55 cents an hour, or 32.7 per cent, below that of New York.

But differences in hourly earnings are not necessarily the same as differences in labor costs. A further element that needs examining is productivity. In a piece-work industry such as this, we must ask whether a differential in hourly earnings is due to the fact that workers in one place can turn out more work per hour than workers in the other.

This is a subject on which hard data are virtually nonexistent. As nearly as anyone in the industry is prepared to say, New York City workers have been more productive than those outside urban centers. The pace of life in a major city is itself a factor which has a bearing on the speed of work. Furthermore, the work force in New York contains a higher proportion of males, who generally have the stamina and motivation to work faster than women.*

* Our guess is that New York City's productivity differential is no longer as great as it was in earlier periods. The shops outside the city are generally larger and more modern in layout and machinery, with more elaborate division of labor, and all this gives them an advantage when they have a large volume of production. At the same time, there has been a decline in average skill in New York, due to the retirement from the trade of skilled male workers and to the fact that those remaining in the industry are older and do not work as fast as they once did. However, the impression of a narrowing pro-

How much of the differential in earnings between New York City and Luzerne County was due to a productivity difference? Our answer is necessarily a guess because of the paucity of information. A study for the year 1934 showed that the weekly output of operators sewing cheap dresses (then wholesaling for $3.75 or under) in suburban areas of New Jersey, Connecticut, and New York State was 10.3 per cent below that of operators in New York City working on comparable dresses.[12] If we assume that a productivity differential of this size also existed between workers in Pennsylvania and New York City in 1955, the labor-cost differential between the two areas can be estimated at about 25 per cent instead of the 32.7 per cent differential in hourly earnings. If we make a further allowance for the cost of training new workers and the greater use of indirect labor in the Wilkes-Barre and Hazleton shops, these shops still have a substantial competitive advantage over New York shops, somewhere in the neighborhood of 20 per cent.

The drive to capture this labor-cost differential is what has pushed garment production into areas, like Luzerne County, having surpluses of cheap female labor. Pennsylvania has a disproportionate share of the nation's areas classified by the United States Department of Labor as "Areas of Substantial Labor Surplus."[13] And it has been precisely in these surplus areas—anthracite counties like Luzerne, Lackawanna, Northumberland, and Schuylkill—that apparel production has grown the fastest. With the decline of anthracite mining, the wives and daughters of unemployed miners have been willing to take work at just about any wages they could get. Meanwhile a developing network of hard-surfaced roads has put their towns within overnight trucking distance of Manhattan's Garment Center. Between 1935 and 1950 the number of women's outerwear plants in the four counties rose from 19 to 212,[14] and there has been continued growth since then.

Even in places not classified as areas of surplus labor, differentials in labor cost have caused growth of garment production—growth

ductivity differential is too tenuous to justify a specific assumption in our calculations.

that was achieved at the expense of even lower-wage industries in the locality. An example is the Allentown-Bethlehem-Easton metropolitan area in Pennsylvania and New Jersey. There, between 1947 and 1954, employment in women's outerwear increased from 3,578 to 6,379, mainly in contract shops working for New York jobbers. As this occurred, employment in the lower-paid undergarment industry dropped from 4,535 to 3,323. Indeed, Seamprufe, Inc., a large producer of women's underwear, opened a plant in Oklahoma because of the difficulty of obtaining additional women workers for its plants in Easton, Pennsylvania. Company officials complained, "We would hire a girl and train her for three months, and someone would try to take her away from us." [15]

Just as the growth of apparel production in Pennsylvania is related to the decline of anthracite mining in that state, so the spread of apparel manufacturing in Massachusetts is related to the decline of the textile industry. The loss of textile production to the South brought down wage rates in Massachusetts, with the result that the apparel industry began to move into New Bedford, Fall River, and other former textile towns.[16] The displaced female textile workers, moreover, had manual skills which were suitable to garment-making; the abandoned textile mills were adaptable to light manufacturing; and many New England communities were willing to provide special incentives to attract the manufacture of garments.[17]

Even within the New York Metropolitan Region, the areas with available labor have scored notable gains in apparel production over the past few decades.* For instance, Paterson, New Jersey, which had been hard hit by the decline of its once-famous silk industry and by the decline of its textile mills, lured garment plants to the area. But the search for cheaper labor within the Region, unlike the search in the small towns and stranded communities of

* Interestingly enough, Connecticut, although very close to New York's Garment Center, is the one state in the vicinity that has not had an appreciable increase in apparel production. This circumstance is due to the absence of a supply of cheap female labor, and this absence in turn is due to the presence of higher-paid machinery industries.

other parts of the country, has been slowed by the marked rise in wages in most of the women's garment industries in the outlying counties of the Region.

In the unit-priced dress industry, in 1947, hourly earnings in the Region outside New York City had been 16 per cent below those in New York City, but by 1955 the discrepancy had been narrowed to only 7 per cent. In underwear and nightwear, a similar discrepancy declined from 19 to 9 per cent between the two dates; in corsets and brassieres from 10 to 3; in children's dresses from 17 to 7; in blouses from 17 to 11. The narrowing was less pronounced in coats, suits, and skirts, and also in knit outerwear, but was still visible.[18] Behind this narrowing lies the fact that pockets of low-wage labor are not so large or numerous in the New York Metropolitan Region as they used to be.

The Region as a whole, therefore, is exposed to the effective competition of other areas in providing cheap labor. The rate of movement to these other areas might have been even faster if the International Ladies' Garment Workers' Union had not sought in recent years to protect the job opportunities of its members in the older centers by modifying its wage demands. A wage increase for the New York cloakmakers in 1957 was their first in four and a half years, and it was designed solely to cover increases in the cost of living. Similarly, an 8 per cent wage increase received by the New York dressmakers in 1958 was their first in five years. Thus, powerful as the union is, it has been restrained by the mobility of the industry from winning substantial wage increases.

But there is widespread appreciation of the fact that this process cannot go on indefinitely. In the long run, the garment industry in New York will maintain its wage position relative to the other industries in the area or its labor supply will shrink in quantity and quality.

UNIONIZATION

Hand in hand with the employers' search for lower costs has been their desire to escape union control. Since the 1920's this desire has provided a great deal of the motivation for the setting up of garment shops outside New York City. Indeed, the start of the exodus of the dress industry from the city was prompted by the wish of employers not to have to grant a number of expensive concessions that had been won by the union.[19]

We have seen how unionization was first established in the outerwear branches in Manhattan in the years prior to the First World War. In the 1920's those who sought to escape union control moved across the rivers into Brooklyn and New Jersey, and northward to Westchester. In time, the union caught up with them, and so they went further—to Pennsylvania and Massachusetts. Today these areas, too, are largely unionized, and the recent movement in undergarments, dozen-priced dresses, and children's outerwear has been to the South.

Although the International Ladies' Garment Workers' Union seems to organize the unorganized eventually, the high turnover of firms and the mobility of the industry mean that organizing is never completed. They also mean that the union, despite the fact that it is one of the most active in the field of organizing, is on a sort of treadmill. As the industry grows in a new era, the union follows it and eventually organizes there, only to find that shops have now sprung up in another area. As a result, the women's and children's apparel industries are no better organized today than they were before World War II. The union has grown, of course, but little faster than the industries themselves. According to a report by the ILGWU General Executive Board, the average monthly employment of production workers in the branches of the women's and children's apparel industries which we are considering was 324,300 in 1939, and the reported membership of the ILGWU for these branches at the end of that year was 221,341, or 68.3 per cent of the production workers.[20] In 1955 the average monthly employment of production workers had risen to 531,400, and union membership at year-end was 367,237, or 69.1 per cent of the production workers.[21]

Branch-by-branch comparisons are not possible, because union membership is not always reported on the same industrial basis as government employment data. It is obvious, however, that the branches that were the most poorly organized in 1939 are much better organized today. Between 1939 and 1955, ILGWU membership in the corset-and-brassiere industry, expressed as a percentage of production workers in that industry, rose from 25.1 to 47.0; in knit outerwear, the increase was from 29.4 to 49.5.[22] At the same time, those branches of the industry, such as dresses, that were very highly organized in 1939, have become somewhat less so, though still well organized, as the industry has moved out of New York and other metropolitan centers.

The ILGWU continues to be more successful in organizing in and around New York than elsewhere. Nearly one-half of the increases in its membership between 1939 and 1953 took place in New York State; [23] yet that state experienced less than one-third of the employment increase in women's and children's apparel in that period. The ILGWU has run into its most formidable opposition in its attempts to unionize the rapidly growing apparel industry in the South. In this area it has met roadblocks of right-to-work laws, organizer licensing requirements, town fees for signing up new members, and outright physical violence by local law enforcement agencies.

Even if organizing in an area meets with success, this does not bring an equalizing of conditions with those in territories that were unionized earlier. Garment production enters new areas because of unequal labor conditions, and an equalization would end the pulling power of the new area. The problem as seen from the national union organization's viewpoint is to raise wage standards in the newly-organized areas as fast as it can, but not so fast as to drive producers into unorganized territory. But the ILGWU is basically a federation of largely autonomous entities.[24] It is composed of more than 500 locals, and in addition there are joint boards and various regional departments. The local unions in each area set their own collective bargaining policies with little central direction from the

International Office. Moreover, the union's position with regard to technological change is not the same everywhere, and employers in New York encounter greater opposition to the introduction of special machinery and of section work. An employer, therefore, may seek out-of-town production not in order to escape the union altogether but only to escape the rigid controls enforced in New York City.

Even if the union were subject to rigid central control, the leadership would find it extremely difficult to eliminate the labor-cost differentials which we have already discussed, for the major explanation of these differentials lies in the nature of the industry—small-scale, fiercely competitive, and highly mobile. These characteristics tend to keep market forces dominant.

Contract shops working for New York City jobbers come under the same collective agreement, no matter where the shops are located. Theoretically, therefore, wage rates should be the same throughout the New York production area. But even within that area, the rates have traditionally been enforced unequally. The dress industry is characterized by complicated piece-rate structures, and the degree of enforcement of union standards can be more crucial in determining actual wages than the contractual rates. Workers in areas like Wilkes-Barre and Hazleton do not normally receive "settlement sheets" similar to those received by workers in the New York Metropolitan Region, informing them exactly how much they are to be paid for each operation. This permits the employers to pay workers much less than workers in New York receive for similar work. In 1956 the general manager of the Dress Joint Board of the ILGWU explained:

Last year our members in the metropolitan dress market produced 97,000,000 dresses in over 50,000 different styles. These dresses were produced in over 2,000 shops in 248 different communities, in addition to New York City, some of them hundreds of miles from each other. *Under our collective agreements, prices are settled with the jobber, but obviously enforcement is possible only where the dresses are made.*[25]

A strike erupted in the unit-priced dress industry in March 1958 on precisely this issue of equal enforcement of the collective agreement in all areas. Union officials stated that the policing features of the contract were as important as the economic demands put forth. In the settlement of the strike, the union won most of its demands, including those concerning equal enforcement of contract terms. At the present writing it appears that this settlement should tend to reduce differentials among areas, but it is highly doubtful that it will eliminate them.

The long-run prospect is that unionization will spread into the nonunion areas, simply because there will eventually be no virgin territories industrially to which the garment shops will be able to flee. But the ferocity of opposition that the union has encountered—particularly in the South—and the continued existence of labor surplus areas indicate that for some time to come, inter-regional differentials in labor costs will remain and will continue to affect location.

Imports of apparel

Differentials in labor costs are not only inter-regional but also *international*. Another aspect of cheap labor that must be briefly considered here is competition from abroad. Imported garments, especially blouses, sweaters, and scarfs from Japan, have occasioned much talk and some excitement in recent years. Compared with total domestic production the imports have been small; but in the mid-1950's, imports of these products showed spectacular increases. For example, between 1954 and 1957, the value of wool sweaters from Japan rose from $336,000 to $7,058,000, and between 1954 and 1956, Japanese shipments of men's and women's cotton shirts and blouses rose from 180,567 dozen to 1,189,925 dozen.[26]

The International Ladies' Garment Workers' Union has reacted vigorously. In October 1955 it won a case against several New York firms which were importing blouses from Japan. The industry's Impartial Chairman ruled that the firms had violated a contract clause prohibiting the purchase of nonunion products, and he awarded compensation of $65,000 to the union. At the same time, employer associations sought governmental assistance to curb imports. The Japanese, faced with American opposition to increased shipments of both textile and apparel items, instituted "voluntary" quotas, which eased the situation but did not wholly satisfy anybody. The ILGWU, moreover, has charged that the curbs imposed by Japan on shipments of garments to American markets have merely shifted the focus to Hong Kong. The union has urged the United States government to set limits on the importation of apparel from Hong Kong.

The somewhat higher levels of garment imports injected another form of low-wage competition into the apparel industry, since Japanese and Hong Kong wage rates are even lower than those in the South or Puerto Rico. This additional competition affected some firms in the New York Metropolitan Region and caused some readjustments. But it probably affected the Region less than the rest of the country. Although there have been signs of an upgrading of imports, the imports have been mainly of inexpensive merchandise, most of which the Region had already lost to lower-wage areas in the United States.

NEW YORK'S GROWING LABOR SHORTAGE

The movement of garment production from New York is related not only to labor-cost differentials and the other labor factors we have discussed, but also to changes that have taken place in the labor force of the metropolis.

In our review of the period up to the 1920's, we saw that the city rose to dominance in the manufacture of women's and children's apparel because of immigration; that the end of free immigration in the 1920's cut off this supply; and that American-born children of immigrants were generally unwilling to become sewing machine operators or hand sewers.

In 1919, Jewish workers constituted 75 per cent of the labor force of the New York coat-and-suit industry, but today they constitute less than 50 per cent. The percentage drops steadily, and will continue to do so, because most of the remaining Jewish workers are of advanced age. Union records show that in January 1958 the predominantly Jewish locals of the New York Cloak Joint Board had the following percentages of members over 60 years old:

Local 117 (sewing machine operators), 50.6 per cent.

Local 9 (hand finishers), 31.5 per cent.

Local 35 (pressers), 42.0 per cent.

The picture in New York's unit-priced dress industry is similar. In 1933 Local 22 (dressmakers) had 30,300 members, 70.5 per cent of whom were Jewish. Twenty years later, when the membership was 25,837, only 50 per cent were Jewish.[27] The decrease in Jewish workers is important because of their skills. Their numbers were continually supplemented by the entry of Italian workers; but in more recent years, there have been strong indications that American-born Italians are spurning the garment industry for occupations with higher pay and prestige.

With the decline of the older ethnic groups that had been associated with the garment industry, there was a need for new elements to enter. Negroes and Latin Americans, mainly Puerto Ricans, have been the two groups which have provided the new manpower—that is, womanpower—for the industry. Being less skilled, they made their entrance in the more standardized sections of apparel. The undergarment trade offers an illuminating picture: young Jewish women were the largest group until the middle 1920's, after which Italian women became most numerous. With the introduction of silks and rayons the need for still more workers brought Negro women into the shops, and by the early 1930's, they comprised 15 per cent of the undergarment labor force. The Negroes first started in lesser skilled jobs, but then moved up the ladder.[28] In the 1940's Puerto Ricans became the major new source of needed manpower.

A similar picture is observable in skirt shops. Again, the original workers were Jewish immigrants from eastern Europe. They were later supplemented by Italians and Spanish-speaking Sephardic Jews from the Mediterranean area. With the great expansion of the skirt industry after World War II as a result of the trend toward "separates" and casual attire, the new workers were to some extent Negro women, but mostly Puerto Ricans and other Latin Americans, who now make up about half the industry's labor force. It was found that, although Puerto Ricans, as the least skilled, were concentrated in the cheapest lines of skirt production, they were able to move both to higher skills and to better price lines.[29]

In the unit-priced dress industry in 1934, Negroes and Latin Americans together made up only one out of every five new workers joining the Dressmakers' Union, but by 1948 they made up half the new members. Here, too, they were largely to be found in the less skilled, lower-paid crafts and in shops making the lower price lines, and in this industry their advancement to higher skills was not proceeding very rapidly.[30] In the higher-skilled coat-and-suit industry the new ethnic groups have hardly made an appearance.

In short, Negro and Puerto Rican women, who are on the lower rungs of the city's economic ladder, have become important in the New York garment industry, but they work mainly in the more standardized branches, and with few exceptions, unlike the Jewish and Italian men of earlier days, they do not become highly skilled tailor-system workers on dresses or "cloaks." As a result, a shortage of skilled sewing machine operators is developing.

The cloak and dress industries are the ones most affected by this shortage, and the union has been concerned with the problem. One attempted solution has been to hold evening classes at the city's High School of Fashion Industries to train section-system sewing machine operators to become skilled full-garment operators. The training of a few dozen women, however, can hardly solve this problem. When people in the industry are confronted with the question of how the industry hopes to meet the shortage of skilled production workers, they generally shrug their shoulders, saying they simply don't know. Some express the opinion that section work will have to spread into higher price lines, with the few remaining skilled workers concentrating in the highest price lines only. Indeed, section work has already made its appearance in the medium price lines, and as this continues to occur, production in the medium price lines can be located further from Manhattan.

The problem of attracting labor in New York is not simply an economic one that could be solved solely by higher wages, but a sociological one as well. American-born men apparently cannot easily visualize themselves at a sewing machine. Furthermore, even high hourly earnings—for example, $2.96 average for male tailor-system sewing machine operators in New York City's coat-and-suit industry (excluding skirts) in February 1957 [31]—do not compensate, at least psychologically, for the uncertainty that pervades seasonal trades. Women entering the industry do not generally plan to spend their lives working in it; so they prefer the immediate earnings of section-system sewing machine operating to undergoing an arduous apprenticeship to become skilled tailor-system operators, even though such training could mean higher earnings in future years.

Even in shops producing cheaper price lines and more standardized products, operating on a section basis and using relatively unskilled workers, New York employers find great difficulty in getting labor at prevailing wages. The labor that is available for the cheaper price lines transfers to higher price lines, where more money can be earned, as soon as workers acquire sufficient skill. Yet, in view of the lower wages being paid in other localities, New York employers in the cheaper price lines cannot easily raise wages.

Negroes and Puerto Ricans, the major new entrants into the women's and children's apparel industries, obviously are not entering in sufficient number to meet all the demand for labor. The question arises: Since even the low-paying branches of apparel in New York offer about $1.60 an hour in average earnings, why don't more Puerto Ricans come to New York to take advantage of these job opportunities, which are better than they can find elsewhere? The answer seems to lie in the fact that the garment industry attracts almost exclusively women workers, while the crucial factor in migration is job opportunities for men. Job opportunities in New York for Puerto Rican men, however, are much more limited. Indeed, Harrison E. Salisbury, in a series of articles on New York's juvenile delinquency problems, quoted one worker in a Puerto Rican residential area who linked the pattern of broken families among Puerto Ricans to the loss of status on the part of the husband when he fails to do as well economically as his wife who goes to work in the needle trades.[32]

Another significant fact about Puerto Rican migration to the continental United States is that New York City's share of it has skidded from 95 per cent to 65 per cent between 1946 and 1956. Table 14 shows the trend. It is generally supposed that the net annual

Table 14 Puerto Rican Migration to New York City, 1946–1956

Year	Net out-migration from Puerto Rico	Estimated net migration to New York City	Migrants to New York City as percentage of net out-migration
1946	39,911	37,900	95
1947	24,551	23,300	95
1948	32,775	29,500	90
1949	25,698	23,100	90
1950	34,703	29,500	85
1951	52,899	42,300	80
1952	59,103	45,500	77
1953	69,124	51,800	75
1954	21,531	16,100	75
1955	45,464	31,600	70
1956	52,315	33,900	65

Source: *Bulletin,* Dept. of City Planning, New York City, February 1957.

migration of Puerto Ricans to New York City in the next few years will be in the neighborhood of 30 or 40 thousand.[33] The opening of new job opportunities in other areas explains the steadily declining proportion of migrants settling in New York City.[34] Recent reports indicate continued growth of Puerto Rican communities elsewhere in the New York Metropolitan Region, especially in the industrial sections of New Jersey.[35] Other reports show that Puerto

Ricans have shifted west, to places like Gary, Indiana, where the men find jobs in the steel mills.[36]

Women do not enter the apparel industries in sufficient numbers because many of them can obtain higher-paying jobs, or at least jobs with greater prestige, in other industries. Women in general prefer the higher status of white-collar work to that of factory labor, and as New York becomes more and more a city of office workers, this preference subtracts from the potential labor supply of the apparel industries. One can probably assume that as Negroes and Puerto Ricans become better integrated into New York City life, they will increasingly forsake the garment industry for alternative types of employment.

FACTORY SPACE AND TRANSPORTATION

The supply and cost of labor, despite their great importance in apparel production, are not the only factors which have a bearing on the location of the industry. Others include factory space, transportation, racketeering, and taxes.

THE AVAILABILITY OF LOFT SPACE

While the existence of low-wage areas elsewhere exerts a pull to attract garment production out of New York City, the shortage of manufacturing loft space in or near the Garment Center acts to push it out.

The apparel industries have not gone in heavily for specially designed plants to meet their own needs, but have been able to adapt just about any sort of structure to their purposes. The small size of the average shop meant that only a couple of thousand square feet would do. Before the turn of the century, women's garment production was concentrated in the tenement buildings of the old East Side. The tenement laws drove it out of residential buildings, and it moved into loft buildings a little farther uptown. The production of women's outerwear continued to edge up Broadway until it found its present famous headquarters more than thirty years ago. Zoning regulations have helped keep it locked where it is. The establishments in the Garment Center, shops and showrooms alike, are piled together in multi-story loft structures that look, on the outside, more like office buildings than industrial plants. The manufacturing space is usually fragmented, running about 2,000 to 4,000 square feet per establishment.

The growth of section work in recent decades, however, has increased the size of the average shop in the apparel industries. A tailor-system shop, in which each sewing machine operator works on the whole garment, can be of any size, but a modern section-system shop, with its elaborate division of the sewing operations, requires 6,000 square feet or more. It is almost impossible to obtain a loft of that size in or near the Garment Center.

The differences in size among coat-and-suit shops are illustrated by a survey conducted by the New York Cloak Joint Board, ILGWU, in the spring of 1953. The survey covered an area roughly equivalent to the New York Metropolitan Region. It found that, although a majority of the production units operated on the tailor system, they employed a minority of the workers. The average tailor-system shop had a work force of 31 as against 49 in the average section-system shop.[37] The location of the different types of shops is also illuminating. Of the tailor-system shops, 79.2 per cent were located in Manhattan, 17.9 per cent in Brooklyn, and only 2.9 per cent "out of town." A reverse order held true for the section-system shops: only 7.5 per cent were in Manhattan, 39.0 per cent were in Brooklyn, and 53.5 per cent were out of town. The average size of section-system shops also varied by location: for example, 41 workers in Brooklyn, 56 workers out of town.

The resistance of the workers and the union to section work is only part of the explanation for the scarcity of section-work shops in Manhattan. Even in those apparel industries, such as undergarments, in which there has been no opposition to section work, firms that wish to remain in New York City find it difficult to do so because of the absence of suitable manufacturing space. When they do find proper loft space it is far from the Garment Center.

The Garment Center, offering unparalleled advantages in merchandising and all sorts of external economies for manufacturing, has become increasingly short of factory space. No new manufacturing loft buildings have gone up in the Garment Center for over twenty years (for very good financial reasons which we shall discuss presently), though some office buildings containing showrooms have been erected. Furthermore, in existing buildings, when a loft is vacated by a manufacturing firm it is often subdivided into offices, cutting rooms, and showrooms for a number of new jobbers.

Similar circumstances in the millinery industry led the United Hatters, Cap and Millinery Workers' Union in 1958 to buy a loft building in the millinery district, which is adjacent to the Garment Center. A major factor in this union's decision to become a landlord in its own industry was the danger that the loft buildings would be converted to other uses, forcing producers of women's hats to shift their operations to other areas.[38]

The scarcity of suitable manufacturing loft space is by no means confined to the Garment Center. It is felt in other Manhattan districts where garment production is located. Of all the floor space in loft buildings in the central business district of Manhattan (that is, Manhattan up to about 59th Street), 44 per cent is in buildings erected before 1914, 47 per cent is in buildings erected between 1914 and 1929, and only 9 per cent is in buildings newer than 1929.[39] As a result, obsolescence and inefficiency, particularly with regard to facilities for the loading and handling of goods, characterize many of the loft buildings.

The showrooms of the undergarment industry are located along Madison Avenue in the East Twenties and Thirties, but the undergarment manufacturing shops are along lower Broadway, in the old loft district, which has been termed by New York's Fire Commissioner "a fourth-rate, shabby, deplorable neighborhood."[40] The age of the loft structures makes much of that district unsafe. This was dramatically emphasized by the death of 24 persons in a fire in one old loft building in March 1958. Most of the victims were women's undergarment workers. This disaster, which recalled the Triangle Fire of half a century ago, prompted the Fire Department to intensify its building-by-building inspection in that district, and to extend it northward to the Garment Center itself, where serious fire hazards were discovered.

Moreover, in the area south of 14th Street there has been a net decrease in loft space as blocks devoted to light manufacturing have been wiped out to make way for new housing developments. For Manhattan as a whole, in the judgment of New York City's Department of City Planning, the existing inventory of loft space will drop still further as buildings are razed to make way for new office and commercial facilities, private and public housing projects, highway construction, and other public improvements.[41]

Because of the lack of space in Manhattan, garment shops in growing numbers are being set up in Brooklyn and Queens. A number of corset manufacturers who wish to maintain plants in New York City have relocated their production from Manhattan to Long Island City, where they can find adequate loft space. A modern section-work plant for the production of women's popular-priced coats, with 50,000 square feet of working space set in a two-acre plot, was opened in Long Island City in the fall of 1958. The president of the firm explained, "When manufacturing under the expensive, cramped and archaic conditions prevalent amid the canyons of Seventh Avenue became economically unrealistic, we made a careful study of what other leading companies outside the needle trades had accomplished through relocation and modernization."[42] Similarly, one of the most modern shops in the underwear-and-nightwear industry, designed for 500 employees, was recently opened in the Ozone Park section of the borough of Queens. The firm had formerly operated five smaller shops in Manhattan. Since its new location is near a subway station, it was able to retain most of its work force, and it was further aided in this respect by the fact that many of the Italian workers, and a growing number of the Negroes, live in Queens.

But even when firms can find manufacturing space in Queens or some other borough, or elsewhere in the New York Metropolitan Region, the lack of such space in or near the Garment Center often acts as a stimulus to push them out of the city, and even out of the Region. Leaving Manhattan involves the loss of some of the external economies mentioned in Chapter 2, the suppliers and services that are clustered there. In order to offset these losses, the firm that is forced to consider an alternative location is likely to seek one where labor is plentiful and production costs are low. Locations further removed from New York, such as the anthracite region of Pennsylvania for dress production and the South for more standardized goods, generally fit these requirements.

In other words, if a firm has to leave the central business district of Manhattan, it may well decide to get out of the New York Metropolitan Region altogether. Just about any small town can provide facilities for garment manufacturing, because the production process is very readily adaptable to any types of quarters which offer adequate floor space. In the small towns of Texas, for instance, most of the shops were housed in buildings previously used as garages, offices or retail stores. One was in a former funeral home, another in what had been a roller skating rink, and two in old aircraft hangars.[43]

At some price, of course, more space would be available in Manhattan, but garment manufacturers cannot pay that high a price. The analysis is the same as that applied to labor: In view of the alternative uses to which space in Manhattan may be put, its price to garment manufacturers is too high in comparison with the price that they would have to pay for space in other locations where the alternative uses for that space are fewer. For example, space in the Garment Center is bid up in price by textile and apparel design houses and others servicing the garment industry, by manufacturers of high-style goods, and by jobbers in all price lines who need showroom space.

Most studies of the apparel industry have dismissed rent as a locational factor, since it comprises under 2 per cent of the total cost of producing a garment. But in view of the jobber-contractor relationship we must take rent into account. In terms of cost structure, the contractor's position is analogous to that of the producer of an intermediate good, who is selling to the producer of the ultimate good, the jobber. The jobber seeks to have the garment sewed and trimmed at the lowest possible price. The general practice in New York is for him to pay the contractor the cost of direct labor, plus an agreed-upon percentage (usually in the neighborhood of 30 per cent) to cover his overhead and indirect labor costs. The union's settlement of labor costs per garment has not been applied as effectively outside the New York Metropolitan Region as within it, and the common practice elsewhere has been for the jobber and contractor to reach agreement on a flat price per garment, leaving it up to the contractor to allocate the money between labor and his overhead costs. Lower rents are a factor, thus, in permitting the contractor to produce the garment for less.

The tightness of space in Manhattan has induced the New York State legislature to continue rent controls, but, even so, Garment Center rentals are considerably higher than elsewhere. The average annual rental in the Garment Center is about $1.65 per square foot in the manufacturing buildings. If one is willing to leave the Garment Center, he will find much lower rents. A study of newspaper advertisements for manufacturing loft space, subsequently confirmed by leading New York realtors, indicated that in 1956 average annual rents in the New York Metropolitan Region varied as follows: Manhattan outside the Garment Center and other specialized districts, 98 cents per square foot; Queens, 97; Brooklyn and the Bronx, 67; Nassau-Suffolk, 92; Bergen-Essex-Hudson, 78; and Westchester-Fairfield, 73. Outside the Region, rents are still lower. Many communities have industrial development corporations which attract new business to their areas through the offer of subsidized rentals. Even without subsidization, rents in surplus labor areas are generally quite low in comparison with rents elsewhere. In those parts of Pennsylvania that have most strongly attracted the production of unit-priced dresses, space is available at a cost of about 50 cents per square foot.[44]

Not only are rents in Manhattan higher than elsewhere, but also, as has been indicated, suitable space is not readily available, even at these high rents. In order to alleviate the shortage, the Mayor's Committee for World Fashion Center has called for the erection, with the city's help, of a fashion center just below Herald Square, and for the construction of loft buildings in the West Twenties between Broadway and Eighth Avenue.[45] (See map, Chart 1, at the beginning of Chapter 1.)

That part of Manhattan would be a suitable area for shops of apparel manufacturers and contractors. But it is easy to see why private builders have not provided the needed facilities. The cost of erecting a nine-story loft building in that area would require a rental of $3.68 per square foot in order to provide a return that would be attractive to investors.[46] This $3.68 would be about $22,000 a year for the 6,000 square feet that a section-system factory for the production of outerwear would need. At the existing average rental per square foot in the hard-coal area of Pennsylvania, the annual rent bill would be $3,000. The difference in rental cost of $19,000 would be more than a contractor might hope to earn in a year as profit.

Even if a firm could obtain the needed space in the Garment Center at present average rents, its annual rental would be $9,900, and that would be over $5,000 a year more than in northern New Jersey, and $6,900 more than in Pennsylvania. Lower rentals in Pennsylvania's hard-coal area, combined with the lower labor costs there, explain why that district has become so attractive to the producers of lower-priced merchandise in the highly competitive dress industry.

The impact of trucking

But the low-wage, low-rent areas could not have risen as apparel centers without fast transportation linking them to Manhattan. The truck permitted the production process to be carried on in those areas, and the construction of more and better highways has allowed the distance between jobber and contractor to increase over the years. This distance will continue to increase as new expressways are built.

Paradoxically, as the production has moved out, the traffic congestion in the Garment Center has become more intense. Trucks deliver piece goods to the jobber's establishment in the Center, then deliver the cut materials to contractors' shops for sewing, bring the sewed garments back to the jobber's headquarters, and finally cart them to railway terminals or directly to retail stores. The buildings in the Garment Center are ill-equipped to handle this daily flow of goods. Since the majority of them have no facilities for loading and unloading, the garment establishments use the streets as part of their factories and assembly lines, and the trucks lining these streets are an integral part of the operation. Since the average block in the Garment Center contains 475 separate firms, trucks trying to make pick-ups and deliveries clog the streets and create a spectacle that, in the rush seasons at least, has to be seen to be believed.[47] All this means wasted time, which contributes to the cost of doing business in the Garment Center.

Even without wasted time, trucking adds to the cost of garments sewed by out-of-town contractors. But, as long as the additional cost is lower than the differential which these contractors can offer the jobber below the price that Garment Center contractors ask, it is to the jobber's advantage to pay the trucking costs. In the unit-priced dress industry the cost of hauling the fabrics to points in eastern Pennsylvania, and the garments on hangers back to New York, ranges from 4 to 11½ cents per dress, depending on the value of the dress. The slightest labor-cost differential between New York and out-of-town locations clearly compensates for the costs of trucking garments back and forth.

Using New York City contractors outside the Garment Center—for example, in Brooklyn or Queens—also requires trucking because the cut goods and finished garments still have to be transported between the jobber's and contractor's establishments. The difference in cost between trucking to Brooklyn and trucking to Pennsylvania is slight, ranging between half a cent and two cents per dress. The round-trip trucking cost on a dress wholesaling for $8.75 is 7 cents between Manhattan and Brooklyn, and 9 cents between Manhattan and eastern Pennsylvania. Clearly a jobber debating whether to use a contractor in Brooklyn or one farther away will not be greatly influenced by trucking charges.

THE LOCATION OF CUTTING AND SHIPPING

Although New York jobbers seek to have their production carried on in the plants of contractors far removed from the city, they generally retain control over the cutting and shipping operations. By doing their own cutting, jobbers can conserve material and hold their costs down. By returning the garments to their New York headquarters before final shipment to retail stores, the jobbers can coordinate orders, assemble the different parts of a garment, and perform other final operations—or, as in embroidery work, have them performed by special shops in Manhattan. One of the most notable examples of the retention of cutting and shipping in Manhattan is the case of Joseph Love, Inc., which produces three million children's dresses a year in fifteen plants throughout the United States and Puerto Rico. This firm does all its cutting at its central plant in Harlem, sends the cut goods to its subsidiary plants for sewing, and then returns the garments to Harlem where they are pressed, finished, and shipped out against orders.[48]

Even so, despite the strong advantages of a New York cutting room or shipping headquarters, the high cost of space and the traffic congestion in the Garment Center have led to a growing tendency for large firms to transfer these operations out of the city. Increasingly, cutting has been moved to the out-of-town factory where the sewing is performed, and shipping has been moved either to the place of production or to special shipping depots located elsewhere in the Metropolitan Region. Such a depot is maintained by Jonathan Logan, Inc. The dresses of this firm, unless sold to a New York store, need never enter the city, for Logan's highly mechanized distribution center in Hudson County, New Jersey, receives the fabrics from the mills and distributes them to the firm's many factories, and then amasses all the dresses and ships them out to customers.[49] There are other large firms that have their garments shipped directly from their contractors' factories. In order for the contractor to do the shipping, however, he must be prepared to perform the work that special accessory contractors in the Garment Center normally handle. Garments that require much hand-work such as tucking and pleating are almost always returned to New York, since there are few skilled hand sewers in the new production areas.

Racketeering

No study of apparel manufacturing would be quite complete without some assessment of the much-publicized matter of racketeering. The tie between racketeering and location is subtle and obscure, but racketeering, it is fairly clear, does exist and does play a role in the movement of apparel production away from New York. Racketeering is not prevalent throughout all sections of women's and children's apparel production, but is limited mainly to the most highly competitive wage-oriented parts, such as the manufacture of inexpensive unit-priced dresses.

The facts, as nearly as they can be reproduced in this murky aspect of the apparel industry's operations, appear to fall in the following pattern: Some jobbers, in their efforts to survive in the perennial struggle against competition, constantly were in search of nonunion contractor shops, with pay scales outside the reach of the union. In view of the union's strength in Manhattan, their search inevitably led to out-of-town locations. But because some aspects of the jobbers' operations, such as cutting and selling, could not easily be transferred out of New York, a constant flow of trucking had to be maintained between these out-of-town locations and the Garment Center.

In the stormy period of the industry's development, when this pattern of operations began, during the 1920's, relations between the jobbers and organized labor were bitter to the point of desperation. In the highly competitive wage-oriented lines, where small differences in cost were crucial, jobbers were not above protecting their access to the nonunion shops by force; union members were not above meeting force with force. The result was that the provision of trucking offered one handhold for the racketeers. It was a small step from providing the services to requiring that the services be used; and another small step to levying unfair charges on the industry through noncompetitive trucking rates.

The last stage in the process appears to have been a branching-out by racketeering interests into the contracting business. In addition to providing the jobber with the trucking, they also designated the out-of-town contracting shops to which the cut cloth was to be shipped for sewing. This was the situation which led a New York State Acting Commissioner of Investigation, exploring the background of individuals attending an alleged "gangland convention" at Apalachin, New York, to say:

> A significantly large proportion of the Apalachin participants are engaged in what appear on the surface to be legitimate activities connected with the garment industry, namely, as manufacturing wholesalers, contractors, truckers, labor "consultants" and union officials.
>
> They are key figures in the non-union garment industry existing in Pittston, Pennsylvania. Pennsylvania sweat shops, muscle-protected trucking bosses and double-dealing union officials have collaborated to undercut the legitimate unionized garment industry of New York City.[50]

In substance, therefore, the role of the racketeer appears to have been that of accelerating a tendency which would have existed in any case—a flight from an area of high-cost labor to areas of lower-cost labor. What the racketeer added was a considerable ability to disregard the union's economic power either to slow down such moves or to organize labor in out-of-town locations whenever such moves occurred.

The effect of taxes

Among the economic forces that ordinarily have some effect on industrial location, we have one more to consider; this is taxes. When executives of garment firms mention taxes as a locational factor, they seem to be stating an afterthought or rationalization of the move, rather than a cause. Contractors located elsewhere, for instance, have listed the absence of a state personal income tax as an advantage of their new locations, but when pressed as to why they had originally moved, they pointed to the availability of labor rather than to the differing tax structures.

When the New York State legislature considered increasing the maximum unemployment insurance tax rate to 3.5 per cent of payroll, which would have increased the tax bill of the highly seasonal garment industry, representatives of both the employers and unions warned that such action would cause firms to move out of New York State.[51] It is highly doubtful that such an increase in the unemployment insurance tax, by itself, would have led to relocation on the part of garment firms, but added on to the higher labor and rent costs, it might have provided further incentive for firms to transfer production out of New York.

Once the availability of cheap labor and other factors have led a firm to decide to move, differing tax structures may enter into an overall complex that helps to determine exactly *where* to relocate. Many states and communities compete in offering favorable "tax climates." In the movement of apparel production to the South, tax concessions, as such, have been of little import. Other special inducements by local communities, however, have been important—for example, free buildings and special financial assistance.[52]

The movement of the corset-and-brassiere industry to Puerto Rico, on the other hand, has been more influenced by tax considerations. Under the terms of the Commonwealth's economic development program, "Operation Bootstrap," firms are not required to pay income taxes for their first ten years of operation in Puerto Rico, an exemption which is renewable under certain conditions. In 1958 the minimum wage in that industry in Puerto Rico rose to 80 cents an hour, which in effect means 85 cents in most plants because the International Ladies' Garment Workers' Union has won a guarantee of 5 cents above the legal minimum. This brought a great reduction in the wage differential between Puerto Rico and the States and caused the island's attractiveness as a potential area of expansion for corset manufacturers to be seriously questioned by leading New York firms. Tax concessions, and the continued availability of labor, however, may continue to attract production to Puerto Rico, despite what happened to the wage differential.

In sum, aside from the Puerto Rican case, taxes do not appear to play a major role in determining the location of establishments in the women's and children's apparel industries.

SAME DRESS, DIFFERENT LOCATIONS

As a conclusion for this chapter on the relative decline of New York as a production center, it is appropriate to illustrate the decline by comparing the cost of producing a dress in New York with the cost of producing it in the hard-coal area of Pennsylvania.

This is a difficult feat because it is practically impossible to obtain accurate data on the cost of producing garments in different locations. First, there is no such thing as a standard product; a dress of one style is made of different fabric from another and requires different amounts of labor. Second, firms are reluctant to give out precise information regarding costs.

We have seen, however, that by using the average hourly wage data of the Bureau of Labor Statistics and by assuming a productivity advantage in favor of New York City, we may conclude that the labor cost of producing an inexpensive dress in the hard-coal area of Pennsylvania is about 20 per cent below that in New York City. This conclusion gives us a basis for estimating how much it would cost a New York jobber to have a dress style produced by a contractor in Manhattan's Garment Center and to have the same dress made by a contractor in the Wilkes-Barre and Hazleton area of Pennsylvania. For a typical dress sold to retailers for $6.75, the comparison is given in Table 15.

In Manhattan, the direct labor cost of the dress is calculated at $1.19 on the basis of a piece-rate schedule that takes into account the labor time required to sew, press, and finish the garment. To these piece rates, 47 per cent is added to cover wage increases won by the union during the years since the schedule was adopted. It is assumed that the jobber handles the cutting in his own establishment.

Table 15 Contracting Cost of Producing a Typical $6.75 Dress in Manhattan and in the Wilkes-Barre and Hazleton Area

	Manhattan	Wilkes-Barre and Hazleton
Sewing	$ .80	....
Pressing	.30	....
Finishing	.09	....
Base rate	1.19	....
Plus 47% cost-of-living bonus	.56	....
Direct labor cost	1.75	$1.40
Plus 30% contractor's overhead (rent, supervisory help, etc., and profit)	.52½	.42
Total jobber payment	2.27½	1.82
Plus 12½% of direct labor cost, for welfare funds	.22	.17½
Total cost of production	2.49½	1.99½
Trucking	None	.06½
Entire cost of contracting production of dress	$2.49½	$2.06

In this case the jobber will pay the Manhattan contractor a labor price of $1.75. The existence of the 20 per cent differential means that he will pay the Pennsylvania contractor a labor price of only $1.40. Both contractors will receive 30 per cent above the direct labor price, to cover their overhead and profit. This means that the jobber's total payment is $2.27½ to the Manhattan contractor and $1.82 to the Pennsylvania one.

The contractor in Pennsylvania probably has a greater indirect labor cost, because he uses more supervisory and floor help and more special machines than the Manhattan contractor. In this respect he has a greater overhead expense per dress; yet he actually receives less per dress. His ability to make money in these circumstances is due mainly to the fact that he produces a greater volume of dresses and pays lower rent than does the New York contractor.

This comparative cost analysis is not completed, however, for the practice in the garment industry is for the jobber to be responsible for the cost of social security, unemployment insurance, and union retirement and health and welfare plans, equal to about 12½ per cent of direct labor costs.[53] In the case of the Manhattan contractor, this would be 12½ per cent of $1.75, and in that of the Pennsylvania contractor, 12½ per cent of $1.40, bringing the total cost of production to $2.49½ and $1.99½, respectively. A trucking charge of 6½ cents brings the Pennsylvania figure to $2.06. Thus, even if the Manhattan contractor is in the heart of the Garment Center and therefore no trucking is needed, the jobber can still have the dress produced for 43½ cents less by a Pennsylvania contractor, even after paying 6½ cents trucking charge between his establishment and the hard-coal area. That is a differential of 17 per cent.

In practice, this contracting differential would not show up quite so clearly. The labor-cost differential between the two locations might mean that a jobber using a Pennsylvania contractor could price his dress at $5.75, instead of the $6.75 required of a jobber using a New York contractor. Or it might mean that he could use a more expensive fabric than the other jobber. In any case the figures demonstrate an unmistakable competitive advantage for the jobber using a Pennsylvania contractor rather than one in Manhattan.

4
The Future

The evidence built up in the earlier chapters now enables us to identify the major forces tending to make the New York Metropolitan Region dominant in the women's and children's apparel industries and the major forces tending to whittle away this dominance. As a preface to our excursion into the future, let us briefly summarize those forces.

For well over a century New York has been the nation's first city and the center of its fashionable society. This predisposed New York to develop as the nation's style center as well. And its reign over women's apparel production was insured by a historic coincidence: the women's ready-to-wear industry was emerging fast in the nation at the same time that large-scale Jewish immigration from eastern Europe was taking place. The production of women's and children's apparel is primarily labor-oriented, and immigration brought New York a pool of cheap labor as well as a skilled labor supply. This was the "sufficient condition" which made Gotham the major source of clothing for the American woman.

Once the immigrants had helped make New York the capital of production, it became also the chief market, and it could build on its position as a cultural center to take complete charge of the artistic designing of garments. As a result of apparel concentration in the metropolis, many other industries and services ancillary to production and marketing gathered around and became, in turn, a source of economies to garment firms, fastening those firms more tightly to Manhattan.

Nevertheless, after the First World War, new forces began tugging at the fastenings, began pulling and pushing apparel production outward from Manhattan. By then, the transition from home sewing to factory production of ready-to-wear clothing had been about completed, and the spectacular phase of the industry's growth was over. Thus, competition became sharper. At the same time the emergence of the dress as a dominant garment increased the importance of fashion, and injected the element of uncertainty of demand for a firm's products. These developments brought the merchandising function to the fore, and here came the jobber, a merchandiser who designed and sold the garment but who had the sewing done in the shops of contractors.

The jobbing-contracting relationship conferred advantages upon the industry; it permitted specialization and brought needed flexibility to a frightfully uncertain business. For though the total demand for apparel products is predictable, the style factor prevents any one firm from knowing what volume of sales a season will bring, and so the use of contracting relieves each seller of the need to maintain factories large enough to fill maximum orders.

The separation of production from marketing, though useful to the industry, has contributed more heavily than any other development to New York's loss as an apparel manufacturing center. The rise of the motor vehicle facilitated the separation and made it possible for employers to enjoy the best of two worlds: a design headquarters and showroom in the Garment Center where garments were sold in the market place and a production shop in the hinterland where wages were lower.

At first, locations in the New York Metropolitan Region outside Manhattan were the chief gainers. But in time apparel production has tended to plant itself still farther away, in areas where pools of surplus female labor permitted the paying of much lower wages than those prevailing anywhere in the Metropolitan Region. The outward flight of garment production was facilitated, too, by the rise of section work, in which most of the tasks could be performed by people of low skills, so that a scarcity of skilled workers became less of a handicap to the hinterland.

Meanwhile New York has experienced increasing difficulty in

keeping up its own labor supply, both skilled and unskilled. The closing of the gates to immigration in the 1920's had cut off the city's traditional source of skilled garment workers. The newer groups in the local labor force, largely Negro and Puerto Rican women, are rapid workers, but are not as skilled as were the male immigrants earlier associated with the industry. As for a supply of cheap labor, the employers are subjected to competitive pressures—competition within the industry from other areas where wages are lower, and competition within the New York Metropolitan Region from other industries whose wages are higher. Though garment workers in the Region earn more than garment workers elsewhere, their earnings are not high enough to attract an adequate supply of workers into the industry, and many employers have felt obliged to seek sources of cheaper labor elsewhere.

The high cost of loft space in the Garment Center also acts to push apparel production out of New York City, and even out of the Region. A location outside the Garment Center causes a firm to lose some of the external economies available to it there. In order to offset these losses, the firm that is forced to consider an alternative location tends to seek one where labor is more plentiful and costs are lower. And this very often is outside the Region entirely.

Post-war changes in consumer tastes in clothing have also had adverse effects upon the New York Metropolitan Region. New York apparel firms have specialized in the production of higher-priced, more formal garments, requiring fancier sewing, but the trend in consumer preference has been toward cheaper, simpler, more casual attire.

These forces are important in throwing light on the next 25 years. Their net effect has been a relative decline in employment for the New York Metropolitan Region in the women's and children's apparel industries. Concisely, here is what has been going on: The national pie has been growing gradually; the New York Metropolitan Region's share has been declining gradually; and the net effect has been a fairly stable absolute level of employment in the Region.

We believe that the forces for dispersal will continue to operate in the future, and that not only the Region's share of employment but also its actual number of employees will be somewhat lower in 1965 and 1975. But before we can estimate the size of the Region's slice, it is necessary to estimate the size of the whole pie, that is, national employment in the industries concerned. National employment, in turn, will depend upon two things. It will depend, first of all, on aggregate national demand, and secondly on any technological changes which might affect the amount of labor required to satisfy this demand.

The future national demand for women's and children's apparel is crucial to our whole structure. In projecting its size, we can afford to ignore the foreign consumer and focus upon the consump-

Chart 4

Percentage of Disposable Personal Income Spent on Women's and Children's Clothing, United States, 1929–1956

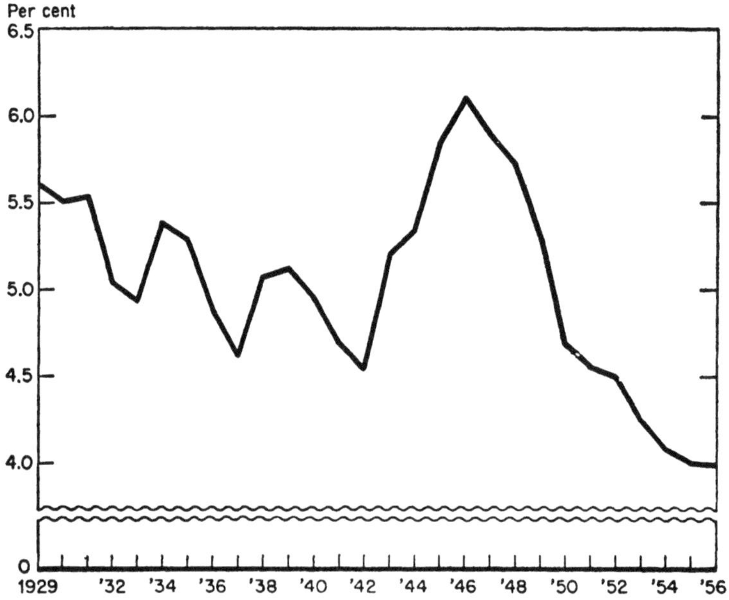

Source: U.S. Department of Commerce.

tion pattern in the United States only, since exports of American-made clothing have always been inconsequential and promise to remain so. We shall begin by seeking an understanding of the trends in national demand up to the present.

Recent trends in national demand

In 1956 the people of the United States spent $11.5 billion for women's and children's clothing.[1] This was more than twice what they had spent in 1929, and it set a new record. Yet these expenditures were increasing more slowly than people's incomes. In fact the expenditures expressed as a percentage of disposable personal income (the amount of money people have to spend from current income after paying their income taxes), have been dropping. This drop is pictured in Chart 4. Between 1929 and 1956, with the exception of the war years, the trend was generally downward, going from 5.6 per cent to less than 4.

Another way of showing the shifts in spending for women's and children's clothing is to present both expenditures and incomes in "real" terms—that is, in dollars of constant purchasing power—and to put both on an appropriate per capita basis. This is done in Chart 5. It gives the relationship between real expenditures "per female" and real disposable personal income per capita in each year from 1929 to 1956.[2]

From 1929 to the middle 1930's, the "per female" expenditures were closely correlated with the per capita income, falling when it fell and bouncing back when it bounced. From 1936 to 1942 expenditures again bore a close relationship to income, but spending on apparel increased more slowly than income, and the line connecting those years on Chart 5 is less steep. In the early stages of World War II, the situation remained unchanged; indeed, union leaders were so worried that they called upon the government to place orders for military uniforms with women's apparel firms in New York.

By 1943, however, the picture had radically altered. Full employment brought a huge expansion of national income. But be-

Chart 5

Expenditures per Female for Women's and Children's Clothing and Per Capita Disposable Personal Income, United States, 1929–1956

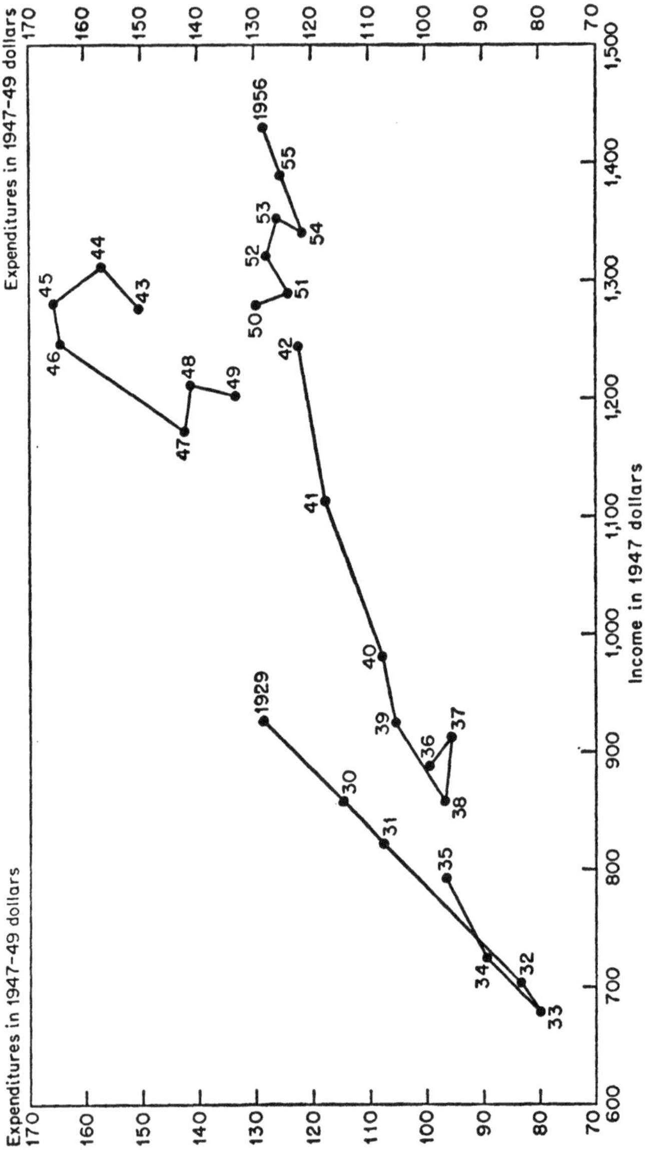

Source: U.S. Department of Commerce.

cause automobiles and other hard goods were unavailable in the wartime period, more was spent on the available items: food, housing, clothing, and entertainment. Thus a great increase occurred in demand for women's and children's clothing, and spending was abnormally high. When the war ended, such spending dropped relative to income, but remained far above that of the early 1940's as long as the flow of durable goods was insufficient to satisfy consumer needs. By the 1950's, however, with durable goods plentiful, the relationship between expenditures and income went back to that of the late 1930's and early 1940's.

Factors influencing demand

Thus we see that while per capita income has been increasing rapidly, the average woman has increased her apparel spending much more slowly. What caused this? To help answer that question we must now examine the major influences upon demand.

Stable per capita demand for products. Some lines of women's and children's apparel are apparently regarded as necessities, falling in the same categories as food and shelter. Therefore, the level of production of these lines is closely related to population. This seems to be particularly true of two major items of women's wear in which the New York Metropolitan Region is dominant: unit-priced dresses and coats. The average American woman shows no tendency to increase her buying of these garments, despite the fact that she has more money to spend. As seen in Table 16, the production per consumer of these garments varied little from year to year between 1943 and 1956.

In the case of women's unit-priced dresses, the average of 2.49 dresses per consumer during that period was identical with the average in 1939. As for coats, the average rose much higher in the 1940's than it had been in 1939, and this higher level has been maintained fairly well.

Demand for women's and misses' suits, however, has been more erratic. At one time, the suit was the chief item of female attire, but by World War I the dress had surpassed it in consumer preference.

During World War II and continuing into the post-war period, the per-female production of suits was generally on the rise, but during the 1950's it was again declining. The up-and-down trend in demand for suits in the last fifteen years is probably due to the operation of a fashion cycle, and over the long run, annual suit production should stay close to the 1943–1956 average of .27 per consumer shown in Table 16.

Table 16 Annual Number of Women's Unit-Priced Dresses, Coats, and Suits Produced per Female over 14 Years Old, 1943–1956

	Unit-priced dresses	Coats	Suits
1943	2.71	.49	.25
1944	2.53	.42	.25
1945	n.a.	n.a.	n.a.
1946	2.51	.41	.33
1947	2.36	.38	.26
1948	2.46	.47	.27
1949	2.78	.46	.30
1950	2.49	.44	.32
1951	2.45	.42	.32
1952	2.54	.46	.29
1953	2.46	.41	.25
1954	2.38	.43	.23
1955	2.46	.43	.26
1956	2.29	.45	.21
Average	2.49	.44	.27

n.a. = not available.

Source: Based upon data of the U.S. Bureau of the Census' *Facts for Industry* series and *Current Population Reports.*

Stability is also observable in the production of children's coats. Such production, measured against the female population under 15 years old, plus the male population under 5 years old,[3] averaged .26 a year for the years 1947 through 1955 and did not stray far from the average. As nearly as we can judge, the production of children's dresses and blouses also seems to be closely correlated with the female population under 15. Corsets, girdles, and other foundation garments are additional items whose production bears a close relation to population. Their production per female over 14 years of age is stable, averaging .85 a year.

On the other hand, certain women's and children's garments appear to be experiencing a trend to greater consumption, and this trend is largely attributable to the growing popularity of casual attire. Skirts, blouses, slacks, and play clothing (beachwear, for example) have made noteworthy per-consumer gains in the post-war period. The middle-income wardrobe of 15 years ago did not contain most of the items that are in it today.[4] However, since the average price of a pair of slacks or a skirt is low as compared to the average price of more formal attire, such as dresses, coats, or suits, the increasing use of these informal garments has not compensated in dollar expenditures for the sluggish demand for the more formal products. And this sluggish demand for "necessity" items, depending more on population than on income, is clearly one reason why spending has not kept pace with income changes.

Family income level. The most important factor affecting the clothing expenditures of different individuals at any point in time is the family income level. Apparel expenditures are higher in higher-income families, but the spending difference is not proportionally as large as the income difference.[5] If family A receives twice as much as family B, it will usually spend more on clothing but not twice as much.

The movement of a family into a higher income bracket apparently does not mean that the family will automatically assume the level of clothing expenditures that such a family would have assumed in former years. We have already seen that in the post-war period, aggregate national expenditures for apparel were influenced by the heavy buying of hard goods when they reappeared in the stores after the war. By the time the family had met its mortgage payments and installments on a car and half a dozen gadgets, it often had little left for the purchase of clothing. This no doubt helps explain why clothes have been declining as a symbol of status in our society. That role is being usurped by durable goods—washers, dryers, and the Cadillac or Jaguar. One writer on the American scene has put it this way:

By and large it is considered improper in our society today to use clothes as a sign of social status or even, in the daily routine, of wealth. Clothes that served a symbolic purpose—the top hat, the frock coat, the starched shirt—have almost completely disappeared, and formal dress has become less common and less formal. Not only do Americans of all classes look much the same when they dress up, they are indistinguishable when they are not dressed up, and that is a larger and larger part of the time. If it takes a sharp eye to tell an original evening gown from an expensive imitation and the expensive imitation from a model that can be purchased by any shop girl, it is quite impossible to tell the rich girl from the poor one when both are wearing blue jeans. . . . Even expensive clothes, though their expensiveness may be obvious to those in the know, do not say to economic inferiors, as clothes used to, "I am different and better than you." [6]

Age. Expenditures for clothing, too, are closely related to the age of the spender. To illustrate, women under 30 spend twice as much for clothing, on the average, as do women over 50.[7] Part of the contrast is explained by the fact that a greater percentage of the women under 30 are in the labor force. This being so, a shift in the population from younger to older people could be expected to militate against spending for clothing. And that is just what has happened in recent decades.

From 1929 to 1956, the proportion of the female population 15 through 29 years old declined from 26.8 per cent to 19.8 per cent, while the 50-and-over age group increased from 16.5 per cent to 23.4 per cent. In actual numbers, between 1943 and 1956, the 15-through-29 age group decreased by one million women. This phenomenon is mainly due to the low birth rates of the depression decade. Medical advances, meanwhile, have lengthened the life span and enlarged the group of women 50 and over.

The average age at which women bear children has decreased, and this fact has also affected clothing purchases, since young mothers tend to spend less on their wardrobes than young childless women. By the early 1950's, 14 per cent of native-born white women 16 through 20 years of age were mothers, compared with only 8.5 per cent in the 1920's; and the proportion of mothers in the 21-through-25 group had gone up from 42 to 56 per cent.[8]

New drastic changes in population structure could affect spending for apparel in the future. Girls born in the 1940's, when the birth rate was high, are now reaching adulthood, and therefore the 15-through-29 age group will make an impressive comeback. On the other hand, the older groups, which have typically spent less on clothing, will also continue to increase, both as a percentage of total female population and in absolute numbers. But the buying habits of women 50 and over may change somewhat. As life expectancy increases, the age when women consider themselves "old" also changes. And besides, more women in the higher age brackets remain in, or return to, the labor force. Thus, even the seemingly adverse demographic trend of an increasing number of older women may be countered by changes in the tastes of these women.

The net effect of these demographic shifts may be favorable to apparel sales, but probably not favorable enough to cause a radical transformation of the spending pattern.

Place of residence. Clothing expenditures are also influenced by where one lives. Families in the North generally spend more on clothing than do families at the same income level in the South and West. Californians, for example, spend less because of the prevalence of informal dress, the milder climate, and the fact that a larger proportion are new home owners. For most income levels, spending is higher in large cities than in suburbs. A notable exception is the suburbanite family earning over $7,500 in the North, which outspends the city family.[9] Accordingly, both the westward trend and the suburban trend of the post-war period have adversely affected clothing expenditures. Those two trends are likely to persist, and should mean a continued depressing influence on the American woman's buying of apparel.

Occupation. On the other hand, a force that should tend to increase spending on clothing in the future is the continued growth of the white-collar labor force. White-collar employees are increasing faster than blue-collar ones. Families of clerical and sales workers spend more on women's and children's clothing than do the families of skilled, semiskilled, and unskilled wage earners, even though the skilled workers have higher incomes than the clerical and sales workers.[10]

The occupational status of women themselves also has an important bearing upon their apparel purchases. Of women under 30 years old, those employed outside the home spend one and one-half times as much on clothing as do the housewives.[11] The trend with regard to working women will tend to boost spending on clothing. There has been a steady accretion in the number of women in the labor force. In 1957 the figure was 21.5 million; by 1965 it is expected to grow to 26.2 million. At the same time a greater percentage of the women workers are in jobs that require dressing up; for example, they are secretaries, salesladies, executives, professional women. In March 1940, only 45 per cent of employed women were in such jobs, but by April 1956, this had risen to 52 per cent.[12]

Another factor tending to increase clothing expenditures is the rising number of young women attending college. By the late 1950's, the number had exceeded one million, and this figure is expected to double by 1970. College women are good customers for clothing, especially freshmen, who spend more money on apparel than do the older students. A survey conducted by the magazine *Seventeen* (January 1957) indicated that the average freshman spent $361.60 on wearing apparel in 1956.

FUTURE NATIONAL DEMAND AND EMPLOYMENT

What do all these trends suggest concerning the future national demand for women's and children's apparel? Various opinions exist. One prognosticator in 1957 held the view that the post-war relative decline of clothing expenditures would not continue, but that expenditures would keep pace with changes in income over the next five to ten years. He based this prediction on an assumption that the trends to the suburbs and to the West, which depress clothing expenditures, will be outweighed by the relative decline of new home owners, the further increase in family size, the fact that war babies will be moving into higher age brackets, the continual decline in rural population, and the relative increase in white-collar workers.[13]

By 1965 disposable personal income should be about 61 per cent higher than it was in 1954, measured in constant dollars.[14] A 61 per cent increase in expenditures for women's and children's clothing would bring such expenditures to $16.7 billion, in 1954 dollars, by 1965. This writer feels that a more realistic 1965 expenditure forecast is indicated by projecting the trend in the percentage of disposable personal income going for women's and children's clothing. Projecting this trend would point to an expenditure of $15.5 billion in 1965, that is, 49 per cent higher than in 1954.[15] A somewhat similar forecast is attained by projecting into the future the relationship between real "per female" expenditures for women's and children's clothing and real per capita disposable income that was evident in the years 1936 through 1942 and 1950 through 1956. (See Chart 5.) Projecting this trend, one arrives at a 1965 expenditure for women's and children's clothing of $15.6 billion at 1954 clothing prices, or about 50 per cent higher than in 1954.[16]

These predictions still tell us nothing, however, about the particular items of apparel that will be bought—a point which is critical in any appraisal of the New York Metropolitan Region's future role in the industry. We have seen that during the last decade the "unit demand" (demand expressed in actual number of garments) for the products of certain apparel industries has been fairly constant per potential consumer, but that the demand for more casual clothing has been growing faster than the number of potential consumers. We assume here that consumer preference among apparel items will continue from 1954 to 1965 approximately as heretofore: that the unit demand for unit-priced dresses, coats and suits, neckwear, corsets and brassieres, children's dresses, and children's coats will increase only as fast as the growth in the number of potential consumers; that the unit demand for blouses, dozen-priced dresses, and underwear and nightwear will grow one and one-half times as rapidly as the number of potential consumers; and that the unit demand for skirts, women's "other outerwear," children's "other outerwear," and knit outerwear will increase twice as fast as the number of potential consumers.

Assuming constant prices, these increases in unit demand would raise the value of these industries' shipments by 28 per cent by 1965. This increase is only about half as great as the anticipated increase in dollar expenditures. The difference between unit demand and dollar expenditures is to be expected, for two reasons:

First, some part of the consumer's dollar will go for consumer upgrading of purchases of apparel. Even though the American woman may put a larger proportion of her clothing budget on relatively inexpensive items like slacks, she may at the same time, as incomes increase, demand slacks of better quality. Indeed, it has already been observed that although consumers have been attracted to lower-priced apparel groups, they prefer better merchandise within those groups.[17] We assume that the upgrading will affect all types of clothing equally. This upgrading should go far toward generating the expenditure level suggested earlier.

Second, we cannot assume constant prices of clothing relative to other goods. Higher relative prices can be expected even aside from changes in quality. Garment prices in 1954 were depressed. By 1965, however, prices should rise more in line with the general price level. Indeed, given the fact that productivity in the garment industry is not expected to increase as fast as productivity in general during the next decade or so, the relative price of apparel should rise even more. And this prospect, when added to the upgrading of consumer purchases, explains why clothing expenditures can be expected to increase by a larger percentage than the actual number of garments bought.

To arrive at the actual projection of 1965 employment, however, some allowance must be made for such increases in productivity as are anticipated. The general sluggishness of productivity change in garment-making was discussed in Chapter 1. The prospect is that future increases will be small, often imperceptible. In most apparel industries, minor improvements in plant layout and machinery may permit productivity to rise at an annual rate of 0.25 per cent. Those industries whose products are more standardized should have slightly greater annual productivity increases: underwear and nightwear, 0.50 per cent; dozen-priced dresses, 0.75 per cent; corsets and brassieres, 0.75 per cent; and knit outerwear, 1.00 per cent. Of course, should there occur the vast technological changes that have been foreseen by some for the apparel industries, there would be much greater increases in productivity. And this would mean smaller future increases in employment than are projected by this writer. In Chapter 1, however, the reasons for doubting such technological changes have been given.

Since the women's and children's apparel industries are already on a 35-hour week, no change in the work week by 1965 will be assumed.

Now we are in a position to combine the prospective increase in demand and the prospective increases in productivity, and thus to project the 1965 employment level for the women's and children's apparel industries in the United States. According to our calculations, 1965 employment should reach about 734,800. The distribution among industries is shown in Table 17.*

The next forecast date is 1975. By that year disposable personal income should reach $600 billion in 1954 dollars, or 46.6 per cent above its 1965 level.[18] By then, however, the nation's personal consumption expenditures for women's and children's clothing ought to have dropped to 3.225 per cent of disposable personal income and thus should amount to $19.4 billion, or 25 per cent more than in 1965.[19]

* The projections for individual industries have somewhat less force as predictions than the aggregate figures. In view of the great importance of fashion changes, unforeseeable switches may occur in consumer tastes and new types of garments may appear. The aggregate figures, however, should bear up better, because if the products of one industry decline in popularity, the products of another will rise to take their place.

Table 17 Projections of United States Employment in the Women's and Children's Apparel Industries, 1965 and 1975

	1954 Employment [a] (thousands)	1965 Demand as per cent of 1954	Annual productivity allowance [b]	1965 Employment (thousands)	1975 Demand as per cent of 1965	1975 Employment (thousands)
Total, all industries	607.1	....	....	734.8	....	834.9
Women's and misses's outerwear	365.0	....	....	445.1	....	509.2
Blouses [c]	44.1	131.5%	0.25%	56.4	120.0%	66.0
Dresses, unit-priced	143.3	121.0	0.25	168.6	116.0	190.7
Dresses, dozen-priced	54.5	131.5	0.75	65.6	120.0	72.6
Coats and suits	74.2	121.0	0.25	87.3	116.0	98.8
Skirts	21.8	142.0	0.25	30.1	124.0	36.4
Neckwear	1.7	121.0	0.25	2.0	120.0	2.3
Other outerwear	25.4	142.0	0.25	35.1	124.0	42.4
Women's and children's undergarments	116.4	....	....	137.8	....	153.5
Underwear, nightwear [c]	74.6	130.0	0.50	91.5	118.8	103.1
Corsets, brassieres [c]	41.8	121.0	0.75	46.3	120.0	50.4
Children's outerwear	78.9	....	....	93.1	....	106.9
Dresses [c]	33.9	116.0	0.25	38.2	116.3	43.3
Coats	14.4	111.0	0.25	15.6	120.0	18.2
Other outerwear	30.6	132.0	0.25	39.3	119.5	45.4
Knit outerwear [c]	46.8	142.0	1.00	58.8	124.0	65.3

[a] Source of this column: U.S. *1954 Census of Manufactures*. Total includes 7,020 employees in Puerto Rico (see footnote c).

[b] These productivity increases are assumed not only for 1954–1965 but also for 1965–1975.

[c] In this industry, which has some plants in Puerto Rico producing for the continental U.S., employees in Puerto Rico are included in the 1954 employment (1,074 in blouses, 1,163 in underwear and nightwear, 2,953 in corsets and brassieres, about 1,400 in children's dresses, and 430 in knit outerwear) and are also included in the projections.

The continued rise in income after 1965 should lead consumers to increase their per capita consumption of all apparel items except unit-priced dresses and coats and suits. We will assume that this increase will be 25 per cent for some products and 50 per cent for the others. Those in the "25 per cent" group are blouses, dozen-priced dresses, neckwear, underwear and nightwear, corsets and brassieres, children's dresses, and children's coats. The "50 per cent" products are those for which unit demand presumably will have been increasing the most rapidly up to 1965: skirts, women's "other outerwear," children's "other outerwear," and knit outerwear.

Again we assume no reduction in the work week. (Possible trends in that direction will be discussed in a larger context in a later volume in this series, dealing with industry generally.) We shall, however, assume that the same small annual gains in productivity will continue after 1965, and we shall reduce the 1975 employment figures to allow for those gains.

Accordingly, 1975 employment in the women's and children's apparel industries in the United States should reach 834,900. This too is given in Table 17, which also gives estimates industry by industry.

Since the hazards multiply as one attempts to penetrate further and further into the future, it becomes perilous indeed to try to predict the demand for women's and children's clothing in 1985. But it is a fairly reasonable estimate that from 1975 to 1985, the number of potential consumers for all of the apparel items will increase by about 12 per cent. Let us make a further rough estimate—that increased income will lead demand to grow one and one-third times as fast as the increase in consumers. By then the increase in labor productivity for all these industries, as a unit, may average 0.50 per cent annually, or 5.11 per cent for the ten-year period. On these assumptions, the 1985 employment comes to 919,000.

✓ Future employment in the region

At last we are ready to return to the New York Metropolitan Region. A projection of the Region's employment in the women's and children's apparel industries based solely upon the national trends in apparel is an incomplete exercise, because of the interaction among the Region's industries. At the present writing (1958), the growth of the apparel industries in the Region has been restrained by an inadequate supply of labor, since labor is being bid away by other types of employment. But on the assumption that this restraint does not become a greater problem in the future than it has been in the past, attempts can be made to predict future employment in the Region.

New York's dominance as a sales and fashion center for women's and children's apparel seems firmly established, and the merchandising function can be expected to grow in the Region almost as rapidly as these industries expand in the nation. Indeed, the growth of air travel is likely to buttress New York's dominance as a sales center. If air travel had not appeared, one might have envisaged the development of large regional sales centers at some stage—at the stage, for instance, when the aggregate demand for high-style apparel in an area like the Midwest or the Pacific Coast was big enough to support the showing of a full range of styles. But air travel places a damper on the need for such regional centers; a trip from Kansas City to Chicago is scarcely less burdensome than one from Kansas City to New York. This gives the New York sales center a fresh vitality which otherwise might have been diluted in time.

In the actual production of the garments, however, the Region can be expected to decline, both relatively and absolutely. The present relationship between the costs of production in the New York Metropolitan Region and the newer centers elsewhere will probably not be much altered for quite some time to come. Of course, the sudden emergence of higher-paying industries in Pennsylvania's anthracite region, or in the South, would upset the present cost relationship. So would a rapid decline of other industries within the New York Metropolitan Region which employ large numbers of Negro or Puerto Rican women, such as the industries turning out electronics equipment, toys, and dolls. The labor thus displaced would augment the labor supply in the apparel industries and lower the Region's cost disadvantage as compared to other areas. A change in American immigration policy might also bring the Region more potential workers for garment-making. These eventualities are not out of the question, but none of them can be assumed here.

As mentioned earlier, the Region's 1956 employment in the women's and children's apparel industries was about 260,000 (see Table 7 in Chapter 3). Between 1956 and 1965 a loss of about 15,000 jobs is foreseen, bringing the total down to about 245,000. The projections for the Region are shown in Table 18. The greatest losses will take place in the cheaper and medium-priced lines of those industries in which the Region is still dominant, that is, unit-priced dresses and coats and suits. The absolute decline will be largely limited to Manhattan, but the growth in the rest of the Region will not be sufficient to offset Manhattan's decline.

Clearly, the apparel industries will continue to be a tremendous force in the economic life of Manhattan. The showroom, sales, and central office functions will grow in Manhattan as the apparel industries grow. Similarly, the production of higher-priced merchandise, dependent upon supplies of skilled labor, will remain very strong in the Region, particularly in Manhattan's Garment Center. The production of high-style apparel in the New York area, like its sale, will be benefited on balance by the growth of air transport. In this segment of production, where expensive transport is easily absorbed in the total cost, air freight reduces the pressure to bring production centers closer to the retail market. The introduction of air freight does, of course, open up the possibility of the growth of new national production centers for high-style garments outside the New York area. But the net impact, in our view, will be to support New York's dominant position by shortening its delivery time to the retail outlets of the rest of the nation.

However, the production of more standardized, cheaper merchandise, dependent upon supplies of low-wage female labor, will continue to drop. As the Region's supply of skilled tailor-system

Table 18 Projections of New York Metropolitan Region Employment in the Women's and Children's Apparel Industries, 1965 and 1975

	1965			1975		
	U.S. employment (thousands)	Region's share	Region's employment (thousands)	U.S. employment (thousands)	Region's share	Region's employment (thousands)
Total, all industries	734.8	33.4%	245.7	834.9	27.6%	230.4
Women's and misses' outerwear	445.1	34.8	154.8	509.2	28.8	147.3
Blouses	56.4	26.0	14.7	66.0	22.0	14.5
Dresses, unit-priced	168.6	43.0	72.5	190.7	35.0	68.5
Dresses, dozen-priced	65.6	8.0	5.2	72.6	7.0	5.1
Coats and suits	87.3	44.0	38.4	98.8	36.0	35.6
Skirts	30.1	43.0	12.9	36.4	35.0	12.7
Neckwear	2.0	80.0	1.6	2.3	70.0	1.6
Other outerwear	35.1	27.0	9.5	42.4	22.0	9.3
Women's and children's undergarments	137.8	28.0	38.6	153.5	22.6	34.7
Underwear, nightwear	91.5	25.0	22.9	103.1	19.0	19.6
Corsets, brassieres	46.3	34.0	15.7	50.4	30.0	15.1
Children's outerwear	93.1	32.8	30.5	106.9	25.7	27.5
Dresses	38.2	32.0	12.2	43.3	26.0	11.3
Coats	15.6	47.0	7.3	18.2	39.0	7.1
Other outerwear	39.3	28.0	11.0	45.4	20.0	9.1
Knit outerwear	58.8	37.0	21.8	65.3	32.0	20.9

workers declines, and the remaining skilled workers concentrate in the higher-priced lines, the production of medium-priced merchandise will have to shift to the section-system method of operation, and much of it also will move to the lower-cost areas where workers of less skill can be utilized. At present, garments requiring hand-work either are entirely sewed in the Region or are returned to the Region for the final hand-work, since the newer areas of production have not yet developed skilled hand tailors. In time, however, the workers in those areas can be expected to develop greater skill, and the dependence upon the Region should decline. On the other hand, a small proportion of even the cheapest, most standardized merchandise will continue to be produced within the Region. The Region, with its small, more flexible tailor-system shops, close to the Garment Center showrooms, will be in the best position to fill small orders and those orders requiring very rapid delivery.

As has been stated, the Region's predicted net loss of 15,000 jobs will be due to the fact that the decline in employment in New York City will be greater than the increase in the rest of the Region. By 1965, the city will probably lose about 23,000 jobs while the rest of the Region gains 8,000. The city's loss will continue to be mainly in the lower-priced and medium-priced lines of unit-priced dresses and coats and suits. The loss in dresses will not be to the benefit of the rest of the Region, because lower-priced dress production tends to move outside entirely. But in the case of coats and suits, the rest of the Region should continue to increase its employment. Indeed, the rest of the Region should either hold its own or slightly increase its employment in all the women's and children's apparel industries. As mentioned in the preceding chapter, the increase promises to be especially noticeable on Long Island; in Dutchess and Orange Counties at the north of the Region; and in the New Jersey counties of Hudson, Passaic, and Bergen.

The forces operating until 1965 will probably continue to govern the locational pattern of the women's and children's apparel industries after that date. Garment manufacturers will probably continue to seek out all possible low-wage areas, whether in Puerto Rico, the South, or distressed areas of the Northeast. As a result, the Region will perhaps lose another 15,000 jobs in women's and children's apparel by 1975. Such a decline would bring the total to about 230,000. A detailed picture of the 1965 and 1975 employment projections for the Region is presented in Table 18.

The net loss of 15,000 jobs will again probably be the result of a greater decline of employment in New York City than the gain in the rest of the Region. By 1975, the city stands to lose an additional 22,000 jobs while the rest of the Region gains about 7,000.

Not much can be hazarded about the state of things as far in the future as 1985. But it is our surmise that in those distant years the Region's share of national employment will decline at a somewhat slower rate, with the result that by 1985 its employment in actual numbers should have been stabilized at a level not far below that of 1975—say 225,000.

That would be only 25 per cent of national employment, as contrasted with about 50 per cent in 1947. Nevertheless, one can foresee, assuming no global catastrophes, that the Garment Center, far from becoming a ghost town, will continue as a teeming place of drawing boards and showrooms and trucks and perhaps helicopters or whatever features will characterize the urban life of the future. Buyers still will flock there from all parts of the country. Entrepreneurs still will pursue the "hot number" that may make them rich, and if they fail, they will bide their time and try again. The American woman, revisiting New York in the 1980's, will find Manhattan Island still supreme in the design and merchandising of her clothing.

PRINTING
AND
PUBLISHING

By

W. Eric Gustafson

Appendix C, Classification of Printing and Publishing Industries, begins on page 341.

Appendix D, Additional Printing and Publishing Statistics, begins on page 344.

Notes to Printing and Publishing begin on page 347.

I

A Locational History

The printing and publishing industry is one of the leading employers in the New York Metropolitan Region, supporting more than 160 thousand workers, or about 9 per cent of the Region's manufacturing labor force. For Manhattan, the industry's importance is much greater. It employs over 20 per cent of the manufacturing labor force there, and together with the garment industry accounts for well over half the manufacturing employees on the densely crowded island.

New York's printing and publishing are important in the nation as well. Practically every town which is more than a wide place in the road has a job printer and a newspaper; yet the New York Metropolitan Region, with something less than 10 per cent of the nation's population, has 20 per cent of the nation's employment in printing and publishing, an exceedingly high percentage in view of the wide dispersal of the industry.

But printing and publishing are significant not merely for their quantitative importance. The industry illustrates many of the characteristics of city-oriented industries generally; and the publishing sector, especially, illustrates many of the characteristic attractions of New York for certain types of economic activity.

The industry is divided into a number of sub-industries, which are listed in Table 1. The table gives some notion of their quantitative importance within the industry in the Standard Metropolitan Area, a territory almost equivalent to the New York Metropolitan Region as delineated in this book. It also shows the importance of the New York area within each branch.

Table 1 Printing and Publishing in the United States and the New York Standard Metropolitan Area, 1954

Industry	U.S. employment	N.Y. area employment	Ratio, N.Y. area to U.S.
Printing and publishing, total	804,386	159,206	19.8%
Publishing industries			
Books: publishing and printing	34,704	15,065	43.4
Periodicals	62,428	26,771	42.9
Miscellaneous publishing	18,378	6,059	33.0
Printing industries			
Newspapers	281,799	28,466	10.1
Commercial printing	200,233	31,793	15.9
Lithographing	77,717	13,685	17.6
Greeting cards	21,347	5,353	25.1
Book printing	22,695	5,625	24.8
Ancillary industries			
Bookbinding and related industries	37,304	12,376	33.2
Printing trades service industries, total	47,776	14,010	29.3
Typesetting	14,550	4,700	32.3
Engraving, plate printing	7,510	2,513	33.5
Photoengraving	17,711	4,758	26.9
Electrotyping, stereotyping	8,005	2,038	25.5

Note: For discussion of the industrial classification used and the meaning of the various industries' titles, see Appendix C.

Sources: For United States totals, *1954 Census of Manufactures,* Vol. II, Part 1 (Washington, 1957). For area totals, special Census tabulation reproduced in Appendix D, Table D–1.

Even a superficial glance at this table shows us that we have diverse forces at work within the industry. The publishing branches —periodicals, books, and "miscellaneous" (including maps, music, directories, and so forth)—have quite high concentrations in the New York area. In the second group, which includes newspapers and the actual printing branches, the area's share is low by comparison. In the third group, composed of bookbinding and the printing service trades, the concentration in the New York area is considerably higher than it is in commercial printing. As we shall see later, this fact is partly a reflection of the subcontracting of specialized services in New York—services of the kind which, in Jonesville, would be performed under a single roof in a commercial printing establishment.

Our purpose in this study is to explain New York's dominance in the field of printing and publishing, with a view to making a judgment about whether, and in what form, this dominance will continue. As a start, we will examine the locational history of the industry, focusing on New York.

THE EARLY DEVELOPMENTS

Not until the nineteenth century did the American printing and publishing industry divide itself into distinct branches like job printing, newspapers, magazines, and books. During the first half of that century, says an historian of the book trade:

> In innumerable small towns throughout the country . . . active publishing was being carried on, as it had been throughout the colonial period, by the individual who combined in himself the functions of printer, publisher, newspaper editor, bookbinder, and bookseller. His product was largely confined to almanacs, schoolbooks, chapbooks, and reprints of favorite works of fiction, travel, and adventure, but first and last it formed a very large part of the reading matter of the residents in the small towns themselves and in the farms surrounding them.[1]

These widely scattered and many-sided individuals, usually calling themselves merely "printers," were also the magazine editors of their day, and Frank Luther Mott notes that "All the prominent magazines before 1825 were supported by the contributions of local coteries and by subscriptions drawn chiefly from within a radius of fifty miles." [2]

The leading cities, even before the Revolution, were Boston, New York, and Philadelphia. But they were not, as they are now, suppliers of the national market, except for an isolated book or periodical which achieved national fame. They were large centers of printing and publishing because they served large local concentrations of population. There were many other centers as well. For instance, there were printers who published books over a considerable period of time in places which are now innocent of any book-publishing activity. Reading, Lancaster, and Germantown, Pennsylvania; Brattleboro, Vermont; Lexington, Kentucky; Charleston, South Carolina; and Newport, Rhode Island, are names which reveal how scattered publishing was.[3]

During the nineteenth century, the picture began to change. With better transportation, lower postal rates, and increasing economies for large-scale printing, centers which had previously served only local markets began to serve wider and wider areas, which became national in scope as the century wore on. National markets appeared for books and periodicals. Job and newspaper printing remained primarily local affairs, however, as they are to this day.

As more and more books and periodicals gained a country-wide circulation, certain centers won an amount of national prominence. Boston in particular was dominant intellectually and in a prestige sense until almost the end of the nineteenth century, although it became increasingly less important in terms of numbers of periodicals or numbers of book publishers. Its pre-eminence was due primarily to New England's role as cultural leader of the United States and to the concentration of native writers in New England, a group which filled the columns of the *Atlantic Monthly,* founded in 1857. Indeed, one of the later editors of the *Atlantic,* Bliss Perry, estimated that at the time of its founding, "the majority of the best-known American writers were living within twenty-five miles of the Massachusetts State House."[4]

Publishing of magazines

Before the Civil War, very few magazines had genuinely national circulation, among them *Harper's Monthly, Godey's,* and *Frank Leslie's Illustrated.* Others, we are told, "carried such words as 'National,' 'United States,' and 'North American' in their titles to show an intention of geographical breadth, but were unable to achieve widespread circulation."[5] The periodicals with national circulation were mostly "class" magazines, in trade terminology. Having little general appeal, they were aimed at audiences of gentlefolk. *Harper's* readers as late as the 1890's were (presumably) entertained by such topics as "Street Scenes of India," "The Social Side of Yachting," and "The Young Whist Player's Novitiate."[6]

After the Civil War, periodical publishing took a great leap. The number of periodicals rose from 700 in 1865 to 3,300 in 1885.[7] Behind this tremendous growth there were three principal factors: the rise of advertising, the fall in costs, and the appeal to mass readership.

As transportation became cheaper and swifter due to the growth of the railroad, markets for manufactured products became more and more national in nature. The spread of the market area carried with it the manufacture of pre-packaged goods with brand names, and national advertising hence became both possible and necessary. The magazines were the only national vehicle for this advertising: newspapers, signs, and handbills would hardly suffice.

Periodicals of an entirely different sort sprang up. The older magazines of national circulation, aimed at a class audience, had been relatively high in price. But the rise of national advertising called for periodicals reaching a much wider audience at lower prices. New techniques of printing and of reproducing illustrations made this possible, enabling the new general-circulation magazines to undercut the established quality journals in price and to reach a much broader market. The older magazines had been printed on slow presses of the same sort used for books, at relatively high cost. As a U.S. Census report noted in 1900, however,

> The greatest advance in printing presses . . . was the construction of perfecting presses capable of producing the finest type and cut work as rapidly as though printing newspapers. Such machines, which were an impossibility in 1880, and an experiment in 1890, are now in general use, and are necessary to the production of the large number of inexpensive magazines and newspaper special supplements, profusely illustrated, which have become an important feature of current literature.[8]

New developments in photoengraving allowed the popular magazines to illustrate as lavishly as had the quality journals, but much more cheaply. Theodore DeVinne, one of the pioneers of American printing, said in 1895: "Illustrations that once cost $100, and that required a month of time, can be had for a tenth of the price, and sometimes in one day. The success of the cheaper illustrated magazines is based on the low cost of ordinary illustration." [9]

Costs of paper were also falling dramatically during the second half of the century. Paper from ground wood pulp was first made in this country in 1867, and by the end of the century it had driven rag paper almost out of the market. Sources differ somewhat in their estimates, but the price of paper seems to have fallen by more than two-thirds between the Civil War and the turn of the century.[10]

Finally, magazine editors began to cater more to the public taste. Rather than depend on the esoteric interests of the upper crust, they began to publish articles dealing with topics of the day and with the everyday lives of their readers. Rather than rely on a ready-made audience of gentlefolk, they created their audience by appealing to the interest of the common man.

Publishers began to discover the guiding principle of the modern magazine business. A publisher, after achieving a low production cost by the use of modern equipment, could sell his magazines at even less than this low cost of production in order to stimulate circulation, and then recoup the loss—and much more—by charging advertisers high rates for the privilege of addressing his readership, attracted by popular articles. At the present time, this tendency has reached the point where publishers receive only about one-third of their gross revenues from the sale of the magazine and two-thirds from the sale of advertising space.

While this general expansion of the periodicals industry was going on, New York City became more and more dominant in the field. Most of the principal new periodicals were founded in New York and a large number of others moved to New York from other parts of the country—Boston in particular.[11] By 1885, in the words of Frank Luther Mott:

New York was certainly the magazine center of the nation, with *Harper's, Century, Scribner's, Forum, North American Review,* and *Cosmopolitan* in the eighties; and then, coming along in the nineties to add to the New York list, *Review of Reviews, McClure's, Munsey's,* and *World's Work*. These were only the leaders among the general monthlies; the great metropolis was just as strong in the general weekly field, and even stronger with its class periodicals—industrial, financial, religious and professional.[12]

Before we offer reasons for New York's gain, it will be well to go back and examine the publishing of books.

Publishing of books

The historical pattern of concentration in book publishing follows much the same course as that in periodicals. During the nineteenth century, three parallel developments led to the centralization of book publishing. First, transportation and communications improved, making it substantially easier to serve the whole country from a few centers. Second, the nation became culturally more unified as native authors began to rise to prominence and the national magazines became more prevalent; when it became more appropriate to publish books for a national readership, the larger publishing houses, through national advertising and forces of salesmen, increased their advantage over the old sporadic local publisher. Finally, improvements in the technology of both printing and paper-making made the economies of large-scale production increasingly pronounced in the course of the century, so that it was no longer economical to issue small editions of books for a strictly local market. Boston, New York, and Philadelphia, already important because they served substantial local markets, assumed leadership on the national scene, and of this trio New York steadily drew further ahead.

Not only did the number of books published in New York increase relative to the number published elsewhere, but also the trend was marked by the actual transfer of firms, especially from Boston. For instance, E. P. Dutton & Co. moved from Boston to New York in 1869, T. Y. Crowell Co. in 1876, and Silver, Burdett

& Co. towards the end of the century. The Manhattan trend also "left many once well-known Philadelphia firms without successors. There existed in the seventies a string of flourishing publishing houses, publishers whose names are practically forgotten today."[13] Most of these Philadelphia houses went bankrupt. Several others abandoned general book publishing for one of the various specialties. Still others were absorbed by other firms, some of them in New York.

In prestige, however, Boston maintained the lead long after it lost superiority in the number of books published, because of the immense reputation built up by publishing native American authors during the middle years of the century. The real shift in prestige, as distinct from quantity, came only around the turn of the century, as Alfred Harcourt, one of the great publishers, reported.

When I entered publishing in 1904 with Henry Holt and Company, the center of gravity of American publishing of books in general literature was . . . only just moving from Boston and New England to New York. Chicago and Philadelphia were important centers; but the New England Brahmins—Longfellow, Whittier, Holmes—and their publishers set the standards of taste in literature and publishing practice for the whole country. . . . New York publishers were in general only following the best of the New England and British tradition.

But between 1890 and 1914, a flood of European immigrants came here—for the last decade of that period, at an average of a million a year. New York became a great Jewish city, a great Italian and German city, and it was inevitable that the cultural heritage and emotional life of these races should find expression. While Boston publishers were bringing out sets of Longfellow and Emerson in new bindings, new publishers sprang up in New York, notably Huebsch, Knopf, and Liveright, who began to publish translations of contemporary foreign authors who had broken away from the Victorian point of view. . . . The older New York houses were quick to recognize the new trend, and New York publishing became international and cosmopolitan.[14]

As the twentieth century opened, then, New York City was unquestionably the book-publishing capital of the United States. And this has been true ever since. No general publishing houses (as contrasted with firms publishing textbooks, medical books, or other specialties) have been established outside of New York since 1900—at least none of any importance.

Talent for publishing

Probably the most important historical factor working towards the centralization of periodical and book publishing in the late nineteenth century was New York's concentration of talent. In 1886, Brander Matthews commented in an article occasioned by William Dean Howells' removal from Boston to New York: "That New York is now the center of literary activity can hardly be gainsaid. . . . There are now more literary men living in and around New York than in or near any other city in America." [15] For those days of slow communication, the importance of having author, publisher, and (in the case of periodicals) advertiser in close contact appears obvious. The literary world is a market place like any other, and the costs of marketing are minimized by centralization. A survey among magazine publishers in 1911 confirms the view that this motivation was predominant:

> Editors claim that their offices must be in New York City in order to secure: 1) The best articles from the best authors, who live in New York City and vicinity. 2) The best illustrations from the best artists, who are also to be found in and about New York City. 3) The biggest advertisements from the largest advertisers, which can be secured and made to yield the best results in New York City.[16]

Much the same has often been said of book publishing. When one center began to gain substantially over the others (perhaps for quite accidental reasons from our point of view), the process of centralization snowballed. New authors seeking a market gravitated toward the growing one; new publishers in search of authors did the same.

Separation of printing from publishing

In the nineteenth century and early twentieth, the printing of books and periodicals had to be close to the office of publication, where the reading matter and illustrations were prepared. Problems of design, illustration, and proofreading, requiring continual contact between publisher and printer, made it difficult to separate publishing and printing operations. Therefore, as New York acquired a greater and greater concentration of publishing houses it also acquired the ancillary printing facilities—whether or not under the same management. At the beginning of the century, as Table 2 shows, New York occupied undisputed first place among

Table 2 Principal Cities in Printing and Publishing, 1900

	Employment	Percentage
United States total	243,999	100.0
1. New York City	47,483	19.5
2. Chicago	20,424	8.4
3. Philadelphia	15,066	6.2
4. Boston	10,056	4.1
5. St. Louis	5,954	2.4
6. Cincinnati	4,758	2.0
7. Minneapolis and St. Paul	4,080	1.7
8. San Francisco and Oakland	3,810	1.5
9. Baltimore	3,517	1.4
10. Pittsburgh	2,909	1.2
11. Cleveland	2,601	1.1
Rest of nation	123,341	50.5

Source: *Twelfth Census of the United States, 1900,* Vol. VIII, *Manufactures,* Part II, *States and Territories* (Washington, 1902).

printing and publishing cities in the United States, trailed at a distance by Chicago, whose employment was less than half of New York's. Other cities had very small fractions as compared with New York, and all the cities with anything approaching a large concentration in printing were publishing centers: New York, Chicago, Philadelphia, and Boston.

New York's lofty position as a printing center, however, went hand-in-hand with its status as a high-cost area for printing. New York printers had to compete for skilled labor against high wage levels in other industries in the metropolis. In addition, the concentration of printing labor in New York made the employers peculiarly susceptible to domination by unions. Labor costs in New York printing were the highest east of Chicago, and since labor is the principal cost component with locational significance, this factor was extremely important. Consequently there was a continuing tension between the area's high costs and the practical necessity, early in the century, of having printing done close to the originating office for nationally distributed products, and close to the market for locally distributed ones.

As the twentieth century progressed, New York lost a great deal of the publishers' printing, which was printing for the national market. The publishing offices did not leave, but they began having their printing done elsewhere. First, distance-sensitive postal rates for periodicals were introduced in 1918, creating a cost advantage for the location of printing at the center of the market. Second, population and income, the chief determinants for the distribution of sales of books and periodicals, became much less concentrated on the eastern seaboard, and as a corollary New York became farther and farther removed from the center of the market. Third, modern communications developed and became cheaper. The airplane and cheap teletype service (in conjunction with technological innovations within printing itself) facilitated the separation of publisher and printer.

Specific impetus for the move outward came from labor troubles at the end of World War I. In 1919 there was a substantial strike which led many publishers to have their printing done outside of New York. In 1921 there followed the famous "forty-four-hour strike" during which the power of organized labor was substantially weakened in competitive cities like Chicago, Philadelphia, and Baltimore. These two experiences led many publishers to try printing at locations outside of New York City and its immediate surroundings, and much of the work stayed there.

The most dramatic move occurred in periodical printing. New York lost all of its large periodical printers, and their departure deprived New York of the printing of *McCall's*, *Cosmopolitan*, and many others.

Book printing, too, began to leave the metropolis, going sometimes to other locations in the New York Metropolitan Region but in many cases outside the Region. Behind these moves lay the improvement of inter-city communications as well as another factor not so important in the case of periodical printing: the development of nuclei of skilled printing workers in places where they had not previously existed, a development brought about in part by the generally rising level of skill and education in the semirural work force of the country.

So the printing of books became less centralized, reversing a historical process. But the new arrangement differed from that of colonial times. The *publishing* of the books remained centralized, and the book-printing firms that now set up in small communities were not there to serve the immediate surroundings but were printing books for a national market.

Other kinds of printing, designed for distribution to the local or regional market around New York, also suffered from New York's high cost levels, but in this case there was an additional problem: if copy was sent out of the city for printing, the printed product had to be brought back to the city for distribution or use. The problem was one of both cost and time: even if the transport could be handled cheaply (which it could not at the beginning of the century), the material still had to be delivered in New York by a "time certain."

The introduction of truck transportation, however, made a big difference. The period around the end of World War I saw a tremendous increase in the use of trucking. Truck registrations in the United States rose to the following levels:[17]

1916	250,048
1920	1,107,639
1924	2,176,838

Truck transportation meant that even Baltimore was within overnight trucking distance of New York City and printed matter could be returned to the city on a tight schedule.

So the picture we have is this: printing for the national market began to leave New York after World War I, followed closely by a good bit of printing for the local market. These two trends did not indicate any sudden change in the cost position of other regions as compared with New York, but rather were a reflection of changes in transportation and communications which made the lower cost structures of other regions relevant as they had not been before.

The printing which remained in New York City was of quite a different character, in four respects which generally went hand-in-hand.

(1) It was work which required extensive consultation between the customer and the printer: examination of proofs, conferences on design problems, decisions of various sorts at particular points in the production process.

(2) It was work requiring highly varied inputs and many specialized services, like fancy typography, special binding, and elaborate work on plates.

(3) It was work done in relatively small batches (compared to the run of a book-club selection or of the *Reader's Digest*), and this meant that specialist establishments handling the various stages of the process could do the work more cheaply than a single printer.

(4) It was work for which delivery times frequently had to be quite short. The question is not only of delivery times on the final product (a problem which rapid truck transportation largely solved for out-of-town locations), but also of quick transfer of goods-in-process between various subcontractors on the work: typesetters, engravers, printers, binders, and mailers, in many cases located in different establishments.

The whole problem will be dealt with at greater length later, but this preview indicates that this kind of printing work, the bread and butter of most printers, has to be done in close contact with the purchaser and by a complex of subcontractors who must also be close together.

NET RESULT OF THE TRENDS

As the more mobile varieties of printing moved out of New York (publishers' printing and the simpler sorts of miscellaneous printing), a process was going on which nevertheless brought more printing of certain kinds into the area. New York has become more and more of a national center, with increasing numbers of printing consumers coming into the city: advertising agencies (especially those specializing in direct-mail and poster campaigns), head offices of corporations, legal firms, foundations, wholesalers, national associations of all sizes and varieties. All these contributors to New York's role as a headquarters city are constant purchasers of printing, much of it work which must be done in close conjunction with the originating office and to a fairly tight schedule.

Publishers have also come to New York in increasing numbers. The publishing of both books and periodicals became much more centralized in New York during the first half of the twentieth century, as shown by figures we will present later.

We can discern, then, three currents at work since 1900: (1) the outward movement of much standardized, large-lot printing, especially for the national market, but for the local market as well; (2) the move into New York of printing consumers requiring their printing work done in the immediate neighborhood; (3) the continuing influx of publishers of books and periodicals.

What are the net results of these trends? Tables 3 and 4 give a statistical picture of printing and publishing since the turn of the century. Figures on employment in the New York Metropolitan Region are not available for years prior to 1947, but a good idea of the earlier trend can be obtained from the employment in New York City and (starting in 1929) in a twelve-county area which the government then designated as the New York Industrial Area. In the first quarter of the century nearly all of the Region's printing and publishing jobs were concentrated in these smaller areas; in

Table 3 Employment in Printing and Publishing, 1900–1954

	United States	New York Metropolitan Region	New York Industrial Area [a]	New York City	Ring of NYMR [b]	Ring of Industrial Area [b]
1900	243,999	n.a.	n.a.	47,483	n.a.	n.a.
1919	446,697	n.a.	n.a.	86,591	n.a.	n.a.
1929	560,909	n.a.	115,896	102,046	n.a.	13,860
1939 [c] ...	552,324	n.a.	109,014	77,560 [d]	n.a.	31,454
1947	715,120	164,745	151,620	131,400	33,345	20,220
1954	804,386	167,227	151,962	126,864	40,363	25,098

n.a. = not available.

[a] Composed of the following twelve counties: in New York State, the five counties of New York City and Westchester County; in New Jersey, Bergen, Essex, Hudson, Middlesex, Passaic, and Union Counties. In the early years, the differences among New York City, the Industrial Area, and the New York Metropolitan Region would amount to only a few thousand employees. In 1900, on the basis of figures for other cities within the Industrial Area, the amount of printing and publishing in the Area *outside* New York City was probably slightly under 5 per cent of the city's total. If we included this additional employment, the 1900 Industrial Area would account for about 20.5 per cent of national employment, as compared to 19.5 per cent for New York City, as shown in Table 4. Although the Bureau of the Census no longer used the Industrial Area as an entity in 1947 and 1954, we arrived at the appropriate employment figures for those two years by totaling the employment in the twelve counties.

[b] The counties outside New York City.

[c] For 1939, "Books, Publishing Without Printing" and "Gravure" were treated because of disclosure rules as though concentrated in New York City, although a small fraction was outside.

[d] New York City figures for 1939 omit the borough of Richmond, but this makes little difference.

Source: Compiled from the Census of Manufactures, area volumes and industry volumes, except for the New York Metropolitan Region figures, which are the sum of federal data for the Standard Metropolitan Area and our estimates for the five additional counties of the Region; these estimates are based primarily on county data from state departments of labor.

1900 very little of the industry existed outside the city proper. Bearing this in mind, we conclude that between 1900 and 1954 the Region's share of national employment in printing and publishing stayed close to 20 per cent. (The percentages appear in Table 4.) The expansion of publishing and the inward trend of printing consumers who brought more business to local printers have filled the place of the printing which has left the area for greener pastures.

Table 4 Ratios of Employment in Printing and Publishing, 1900–1954

	N.Y. City to nation	NYMR to nation	N.Y. Industrial Area to nation	N.Y. City to NYMR	N.Y. City to N.Y. Industrial Area
1900	19.5%	n.a.	n.a.	n.a.	n.a.
1919	19.4	n.a.	n.a.	n.a.	n.a.
1929	18.2	n.a.	20.6%	n.a.	88.1%
1939	14.0	n.a.	19.7	n.a.	71.1
1947	18.4	23.0%	21.2	79.8%	86.7
1954	15.8	20.8	18.9	75.9	83.5

n.a. = not available.

Source: Computed from data in Table 3.

It is important to note that this is a constant share of an expanding total. National employment in the industry has more than tripled since the beginning of the century. The employment in the New York Industrial Area, and almost certainly that of the New York Metropolitan Region, has also more than tripled. A stable share of employment in a stagnant industry would not be nearly as significant.

Within the Industrial Area, however, there have been considerable changes. What we shall call here the "ring" of the Industrial Area—that is, the outlying counties in New York and New Jersey—has gained much work which New York City has lost. The city proper has declined from 19.5 to 15.6 per cent of national employment during the half-century, and the Industrial Area's share of the national total has been preserved only by the growth of printing in the ring. As we can see in Table 3, this growth was especially rapid in the depression years. In the decade 1929–1939, employment in the ring more than doubled, while that of the city declined by almost one-fourth. That decade brought New York City down to only 71 per cent of the Industrial Area's employment in printing and publishing. Since then the city has recouped some of its losses, but a much more substantial fraction of the Area's printing work is now done in the outlying counties than was the case at the beginning of the century.

What has happened to other printing cities during this same period? Table 2 showed the largest printing and publishing cities in 1900. Table 5 shows the eleven largest Standard Metropolitan Areas in printing and publishing employment in 1954. The 1954 distribution is not markedly different, but has several interesting features. In contrast to New York's stability, a number of centers have gained as a percentage of the national employment since 1900.

Table 5 Principal Metropolitan Areas in Printing and Publishing, 1954

	Employment	Percentage
United States total	804,386	100.0
1. New York	159,206	19.8
2. Chicago	81,698	10.2
3. Philadelphia	35,099	4.4
4. Los Angeles	26,894	3.3
5. Boston	23,430	2.9
6. San Francisco-Oakland	15,918	2.0
7. Minneapolis-St. Paul	15,659	1.9
8. Detroit	15,267	1.9
9. Cleveland	14,089	1.8
10. St. Louis	12,471	1.6
11. Cincinnati	11,942	1.5
Rest of nation	392,713	48.7

Note: Standard Metropolitan Areas as used by the Census Bureau.

Source: *1954 Census of Manufactures*, Vol. III, *Area Statistics* (Washington, 1957).

Chicago, San Francisco-Oakland, Minneapolis-St. Paul, Detroit, and Cleveland have all gained. These are all cities in the growing areas of the country, the Midwest and the Far West—as are Los Angeles and Detroit, which were not even on the list in 1900. Pittsburgh and Baltimore, both eastern cities, have disappeared from the list by 1954. The rise of some cities and the decline of others represent two related phenomena: the shift of periodical printing and other long-run, nationally-distributed printing to the Midwest, and the growth of centers of printing serving local and regional markets.

What lay behind these shifts and the likelihood of their continuing can be determined only by looking at each of the main categories of printing and publishing.

2

Serving Local Markets

Part of the printing and publishing industry is closely tied to its market; part is not. At one extreme is the newspaper, which by its very nature must plant itself in the midst of its consumers. Next comes job printing, which varies from work which must be practically around the corner from its customers to work which is free to migrate fairly far from its market. We will deal with newspapers and job printing in this chapter. After that we will take up periodicals and books, which are distributed to the national market and which therefore lie at the other end of the printing and publishing spectrum.

NEWSPAPERS

The newspaper has been historically a local industry in the United States. Newspaper offices are dotted all across the face of the country, in almost every city, suburb, and town. The industry is much less dominated by the metropolitan press than in Europe.

Newspapers are tied to their local markets in three ways. The first is by their focus on local news. Second, their revenues are mostly from local advertising. Seventy per cent of the revenues of American newspapers comes from advertising,[1] and the bulk of this advertising is placed by merchants with geographically restricted markets. Table 6 shows that American newspapers derive most of their advertising revenue from local ads, the percentage being higher for small newspapers (presumably in smaller communities) than it is for large ones. Finally, there is the factor of time which ties papers to their local markets in all communities. Papers are in

Table 6 Local and National Advertising Revenue in Daily Newspapers, United States, 1956

	Percentages in newspapers with circulation of:			
	10,000 to 25,000	25,000 to 50,000	50,000 to 100,000	Over 100,000
All advertising	100.0	100.0	100.0	100.0
Local advertising	87.6	82.6	82.6	74.4
National advertising	12.4	17.4	17.4	25.6

Note: These figures are taken from a sample survey. Local advertising means advertising for locally distributed goods and services; national advertising for nationally distributed ones. Local advertising includes all classified advertising, although a small fraction of this really ought to be under national advertising.

Source: *Editor and Publisher*, 90:18 (April 20, 1957).

competition not only with one another but also with radio and television to get news and advertising to their readers as fast as possible.

The result of these compelling factors localizing the press is that newspaper production in any given area does not depend on the area's cost position as compared with other areas, but rather on the extent of the demand in the area.

METROPOLITAN NEWSPAPERS

Newspapers are a big industry in the New York Metropolitan Region, with an aggregate employment only a little less than commercial printing. In 1954 the 341 newspaper establishments in the 17-county Standard Metropolitan Area accounted for 28,466 jobs.[2] Well over half of these employees work on a few large English-language dailies in New York City.

The newspaper picture in the city is considerably enlivened by its foreign-language press, which must be the most varied of all cities in the world. There are 18 dailies in 11 languages, examples being *Al-Hoda* in Arabic, the *Chinese Journal* in Chinese, *Nowy Swiat* in Polish, and the *Jewish Daily Forward* in Yiddish. The combined circulation of the foreign-language dailies is larger than that of the

Herald Tribune or the *Post*. Besides, there are weeklies in almost all of the 11 languages which have dailies, with 14 additional languages represented as well.

The only other important concentration of newspaper employees in the Region is found in Essex County, largely in Newark. The rest, working on both dailies and weeklies, are scattered through all the other counties in the Region.

Metropolitan dailies in the United States are subject to the same localizing pressures that apply to newspapers in general, and New York City's papers are no exception to the rule. Table 7 shows the extent to which the nation's large-city newspapers circulate outside of their metropolitan areas. New York's pattern is not much different from that of other large cities; the city's seven large dailies taken together send only 13.5 per cent of their circulation outside

Table 7 Percentage of Newspaper Circulation Outside the Retail Trade Area for Newspapers in Leading Cities, 1957

City	Daily	Sunday
New York City	13.5	32.5
Times	25.2	42.3
Herald Tribune	23.0	30.6
Mirror	19.5	36.2
Daily News	13.9	33.6
Journal-American	4.0	18.0
World-Telegram and Sun	2.0	....
Post	1.6	2.9
Chicago	14.9	24.9
Los Angeles	10.0	18.0
Philadelphia	6.4	22.7
Detroit	11.0	21.5
Boston	22.8	34.3
San Francisco	13.5	29.7

Note: A city's retail trade area is the rough equivalent of its Standard Metropolitan Area. Cities are ranked in order of the population of their Standard Metropolitan Areas.

Source: Computed from *Market and Newspaper Statistics* (New York: American Association of Advertising Agencies, 1958), Vol. 26.

of the retail trade area.[3] Sunday papers circulate over a much broader territory, as seen in the table. But they are a different commodity from the daily newspaper. In fact, they can more properly be considered as a subspecies of the weekly magazine, since their proportion of spot news is relatively small, and much of the Sunday paper is ready for the presses by Thursday night or even earlier.

The overriding factor in the production of metropolitan daily newspapers is time. A daily must have its papers on the street within a few hours after the completion of the bulk of its copy. For metropolitan papers, the deadlines are especially pressing because of the competition from other newspapers which put out editions at the same time. For instance, a chief selling-point of the late afternoon editions of New York City's *World-Telegram and Sun* is a complete report on the stock market, requiring continuous setting and resetting of type throughout the day by a crew of almost forty men in order to keep up with market changes. When the stock market closes at 3:30, last-minute calculations are made, and by ten minutes of four there are copies on the street with the complete story of the day's market activity. Papers have similar short deadlines for theater and movie reviews, sports results, and so on. Many papers print late news from new plates several times within a few hours.

Therefore metropolitan newspapers, severely limited in time, and serving a population concentrated in a restricted area, must be at the center of the market, which for New York means Manhattan. The daytime population of Manhattan's central business district (south of Central Park) is nearly five times the number of people who make their homes there. In the circulation of morning newspapers, the chief problem is getting the papers out to the points from which people commute to work. Consequently, the morning newspapers have located in midtown Manhattan, close to the Grand Central and Pennsylvania railroad terminals, so that they can speed the papers out to the suburbs. Evening newspapers, however, have remained in lower Manhattan where all newspapers used to be.[4]

Locational trends show every indication that the present pattern of central locations for newspaper offices will continue in the future. The plant of the *Daily News* on East 42nd Street has just undergone a modernization and expansion. The *Times* has started building a new plant on West End Avenue between 62nd and 66th Streets which will ultimately house all its editorial, printing, and distribution operations.

The problem is not just one of the distribution of the printed newspapers, but is also a publishing problem. In order to keep up with the latest news, the papers are closely tied to the center of their news area, again Manhattan. National and international news comes in over the wire services and from the Washington news bureaus of the papers, and thus does not influence location within the city. New York news dominates the content of the papers, however, and the location which minimizes the cost of communication between the reporters and those who edit the news, set it in type, and print it is at the center of the news area.

There are two possible alternatives to location in the heart of the city. The first depends essentially on the structure of the metropolitan area. If the concentration of population in the center of the metropolitan area is low compared to that of the whole area, it may be possible to locate the plant outside of the Central Business District in order to avoid congestion in delivering newspapers to the consumer. In the Boston Standard Metropolitan Area, for instance, only 34 per cent of the population is in the central city, as compared with New York's 67 per cent.[5] Boston's newest plant has been built outside of the downtown area, in large part as a consequence of this dispersion of population. (Boston's papers also serve a much wider area than those of other metropolitan areas, as suggested by Table 7.) But this sort of move is not a possibility for New York for a long time in the future, both because of the concentration of population and the large number of commuters to Manhattan.

Another alternative is also possible, however, one which depends on the size of a paper's circulation in particular areas. If circulation in a given area is sufficient to permit the construction of a branch plant which is large enough to take advantage of the considerable economies of scale in newspaper printing, a newspaper may find it profitable to establish such a plant to handle part of its demand.[6] In New York City some of the large papers—for example, the *Daily News*, the *Journal-American*, and the *World-Telegram*—have established branch printing plants in Brooklyn to serve the heavy concentration of readers there. Having a plant in Brooklyn cuts out the congestion involved in trucking all the Brooklyn copies of the papers across the bridge. Only stereotype mats need be taken across, and stereotype casts, from which the actual printing is done, are made from these mats at the Brooklyn plants.

THE GROWTH OF SUBURBAN PAPERS

The big papers published in the central cities of metropolitan areas have failed to grow as fast as suburban papers. One prime reason is that metropolitan areas have simply become too large for central-city papers to serve the classical newspaper functions in the suburbs.

In recent years, as Table 8 shows, the seven largest central-city papers have barely held their own in the aggregate, and two have actually declined, with only the *Times* increasing substantially. The 44 suburban dailies, on the other hand, have boosted their total circulation rapidly. The situation seems to be the same in other metropolitan areas as well, although detailed figures are lacking.

What is behind this trend toward suburban papers? When population was still fairly well concentrated in Manhattan and the population growth was occurring in the other boroughs, big downtown advertisers like department stores could still reach these expanding populations through central-city newspapers, a number of which began to include special pages containing borough news and advertising. The expansion then was taking place in high-density areas, where transport costs to the central city (both in money and in time) were low enough so that central-city shopping was still feasible.

In recent years, however, metropolitan areas—and New York in particular—have been expanding primarily in low-density suburbs far from the central city. Twenty miles from downtown Manhattan, for instance, the population per square mile is less than 5 per cent as great as in Manhattan. To serve this diffused population,

Table 8 Daily Circulation of New York City and Suburban Newspapers, 1952 and 1957

	1952	1957	Percentage change
New York City, seven-paper total	5,592,500	5,656,590	+1.1
Daily News	2,161,187	2,083,972	−3.6
Mirror	920,838	876,938	−4.8
Journal-American	689,981	698,881	+1.3
Times	507,397	622,843	+22.8
World-Telegram and Sun	566,923	577,891	+1.9
Post	399,081	428,817	+7.5
Herald Tribune	347,093	367,276	+5.8
Suburban dailies	1,798,224	2,276,097	+26.6

Note: Suburban papers are papers published outside New York City but within the New York Metropolitan Region. Circulations are for six-month periods ending March 31.

Source: Research Department, *New York Times.*

suburban shopping centers have appeared. The Metropolitan Region has changed from one market area for large stores—and large advertisers—to many market areas geographically dispersed.

To be sure, some of the shopping centers which straddle a number of local market areas sometimes find it less costly to advertise in a metropolitan paper—even though most of its circulation goes elsewhere—than to advertise in a number of local papers. Nonetheless, suburban papers benefit heavily from shopping center advertising.

Suburban newspapers also hold attractions from the reader's viewpoint, especially the reader who does not commute to Manhattan every day. The metropolitan paper cannot report the PTA meeting or the intricacies of suburban politics with the coverage the suburbanite would like, nor can he get his name easily in the *New York Times*. The suburban papers can perform these functions for the metropolitan area's increasing proportion of suburbanites.

The task is just too small for the metropolitan paper, both from the advertiser's viewpoint and from that of the reader. Parochialism in advertising and news can be handled by special supplements for the high-density boroughs of the city itself, but there are too many suburban neighborhoods for the metropolitan papers to cope with the problem. Unless the metropolitan papers come up with some new approaches toward penetrating suburban neighborhoods, the publishers of the suburban papers will continue, it seems, to have a protected monopoly position.

The story on newspaper location, then, can be summarized as follows: the straitjacket imposed by the newspaper's local focus, its short deadlines, and its clustered consumers makes newspaper printing an extreme case within the printing industry. Because of these factors, the printing of newspapers is not free to migrate to alternative locations to take advantage of lower production costs, nor are papers from low-cost areas any threat to papers in high-cost ones. Indeed, labor cost for newspaper printing is generally higher than for job printing in the same cities, precisely because labor has been able to take advantage of the protected monopoly position of newspapers, while other kinds of printing are more open to outside competition.[7]

JOB PRINTING

As in the case of newspapers, the location of job printing depends principally on the locations of its customers. And New York has by far the greatest gathering of printing customers in the nation. More than any other city, New York is an office center, devoted to paper-work. Besides the usual printing needs generated in urban surroundings, huge amounts of printing are daily consumed in New York by the financial district and Madison Avenue, by the national headquarters of corporations and trade associations.

During the course of the century, as already indicated in the first chapter, much has happened which has freed these customers from the necessity of having their printing done in New York City, but the pressures are still considerable. Job printing, making up the bulk of printing work, varies in respect to those pressures. It ranges from work which resembles newspaper printing in that costs at alternative locations are irrelevant, to work which is free to seek a location hundreds of miles from the location of its geographically clustered customers.

Most job printing falls within the Census classifications of commercial printing and lithographing. It is distinguished from book printing (which has its own Census classification) and periodical printing (which does not, but falls within commercial printing).

Time and consultation

Much job printing is done under tight time schedules, and is in this respect similar to newspapers. Copy for lawyer's briefs, menus, concert programs, advertising campaigns and so on is often not prepared until shortly before the final printed product must be ready. A reader leafing through *Printing News,* New York's trade weekly, is struck by the emphasis on time and deadlines in the trade advertising. Even if the deadline was yesterday, one would get the impression that there are firms perfectly capable of delivering the goods. The job printer, then, must be practically around the corner from his customers for much of his work.

In Manhattan one major printing district is in the City Hall area, serving the nearby financial and legal offices. A second, which sprang up later, is located around the main post office on Eighth Avenue, serving the retail trade concentration in the area and many consumers whose printed materials circulate by mail. More recently, a third district has grown up north of East 42nd Street, near the Grand Central Post Office, to serve midtown and uptown customers, especially in advertising. Relatively small numbers are engaged in printing elsewhere in Manhattan and in the other boroughs of the city.

Further aggravating the deadline problem is the fact that printing is not a manufacturing operation in which the producer can make a standard item and then stockpile it. Each printed product is a unique item in itself, tailored to the detailed specifications of the individual customer. The printer must frequently help to determine the specifications, functioning in part as designer of the final product and consultant on methods and materials. Then, when layout is done, type set, and engravings made, the customer has to correct proofs of text and examine proofs of engravings to check their appearance. For this sort of consultation, it is almost essential to maintain close and continual contact between printer and customer.

We can observe this process by taking an advertising brochure through the stages of production. The customer could be an advertiser or an advertising agency. For simplicity, let us assume that the advertiser is ordering the printing directly, through his own advertising department. After the brochure has been conceived and planned, there results a layout or "dummy" of the brochure, a typewritten sheaf of copy, and preliminary sketches of the illustrations. Specifications for type faces and art techniques may be worked out at this stage, or left to the printer or a subcontractor. The copy goes to the printer who either does his own typesetting or calls on a composition house to do it—or the customer may deal directly with the composition house. In any event, a series of proofs shuttles back and forth among advertiser, printer, and composition house.

The art elements are subcontracted to artists, photographers, retouchers, and letterers. Again, "roughs," corrections, and final art work shuttle back and forth. Then the art elements must be converted into printing plates; this may bring other firms, such as photoengravers, into the picture. When art and type have been combined, proofs of the brochure are submitted, and then new proofs of the corrected work.

On important jobs, there may even be a "press proof" pulled when everything is ready to run, and this proof may be rushed to the advertiser for telephoned approval while the press waits. After the printing, most jobs require some form of finishing, such as folding, binding, laminating, or punching. Some of these operations may be performed by the printer; others go to outside service organizations. Again samples have to be submitted for approval. All through these interrelated processes, the printer and the subcontractors must be in contact with the customer.

Size of establishment

The unstandardized nature of printed products, besides making frequent consultation necessary, also forces job printers to remain small. As we have seen, the job printer's output is a tailored commodity, whose characteristics vary considerably from job to job. When another item has to be produced, the process typically begins again from the ground up. Type must be set anew, plates made, and all the rest. The largest part of the cost—for most printing work the engraving, typesetting, and platemaking—is a fixed cost for each job. It has to be done only once, whether the printer turns out a hundred copies or one hundred thousand. The significance of the point is this: the scale economies available in printing are economies which arise not from the size of the establishment, beyond a relatively small size, but from the average number of copies produced on specific printing jobs. If the jobs are in general short-run to medium-run, as most printing work is, then the size of establishment is limited by the number of these short-run projects which can be conducted under the same roof without running into increasing costs.

The effective limitation is imposed by the costs of supervision. The manufacture of printed products requires close supervision at many stages of the work, with constant attention to specifications and standards. The larger the establishment, the more involved the administrative process necessarily becomes. By all testimony, bureaucracy and red tape grow at a more rapid rate than the size of the establishment.[8]

Large shops are possible only if the demand comes in large pieces or the work is of a highly routine nature, with the specifications of one job very like the next. These conditions are present in the printing of newspapers, books, and greeting cards. Table 9 compares these industries with commercial printing in the average size of establishment as measured by number of employees, and also in the proportion of the employees who work in small shops. (These two measures are, of course, almost opposite sides of the

Table 9 Establishment Size in Branches of Printing, United States, 1954

Industry	Average number of employees per establishment	Percentage in establishments of less than 50 employees
Commercial printing	16.6	42.2
Lithographing	26.6	34.6
Newspapers	32.6	23.2
Book printing	40.0	22.4
Greeting cards	72.6	10.2

Note: For details of industrial classification, see Appendix C.

Source: Computed from data in *1954 Census of Manufactures,* Vol. II, Part 1 (Washington, 1957), Table 4, p. 27A–9, and Table 4, p. 27B–2.

same coin.) We see that the commercial printing category has the smallest establishments, on the average, and the largest proportion of employees in small firms. Lithographing, for which figures are also presented, is similar to commercial printing, but is done by a different process.[9] This process is better adapted to long-run work, a fact which is reflected in the average size of establishment.

SUBCONTRACTORS

The smallness of the average print shop, enforced by the factors we have just discussed, means that many of the operations involved in printing cannot be carried out at an optimal scale in the individual printing establishment. Printers originally handled all stages of the printing operation themselves. Indeed, as we have pointed out earlier, until well into the nineteenth century even publishing and bookselling were typically conducted in the same establishment with printing. As the volume and variety of the printed product have increased, however, specialized establishments have grown up to handle single stages of the work. This has been especially true where technological innovations have brought about a situation in which various stages of the printing process are done most efficiently at a fairly large scale through the use of capital equipment in large, indivisible units. Good examples of these innovations are electrotyping equipment, which makes a single printing surface from a form containing many separate engravings and lines of type, and mechanical binding equipment.

Because of the large investment required, the particular stage may be performed at lowest cost at a size which is out of proportion to the optimal size for the rest of the operation. Then it will tend to be performed in a separate establishment. For example, a single printer operating at his optimal scale (which is determined in large part by the costs of supervision) might not be able to use the services of a binding department organized at *its* optimal scale. If not, his binding will be done most efficiently by an establishment which serves a number of printers.

The problem arises more acutely in printing than in many other industries because of the diversity of the product. It is not simply a question of the optimal size for a single routine operation, but for a highly varied one. Advertising printing, to select one example, demands a wide variety of type faces. Any one of these type faces would only rarely be needed by an individual printing firm, but would be used fairly often in the aggregate by the large number of printing firms which have typesetting done by a trade composition house.

The fact that the printing process takes place in a number of separate stages increases the opportunities for subcontracting. For letterpress (the most common variety of printing), typesetting and photoengraving come first. Type and engravings are then combined, and printing can be done directly from them. When long-run work is to be done, or when the same material is to be run at intervals, electrotypes can be made from the type and the engravings to provide a single durable printing surface. Then comes the actual presswork stage, followed by binding and other finishing operations. Finally, in many cases, the addressing and mailing can be considered a part of the printing operation itself.

If the printing is to be done by lithographing or gravure, the stages are similar to those followed in letterpress work, with the exception that type is never used directly on the press. Instead, a proof is made from the type and photographed in combination with the illustrations to produce an offset or gravure plate. The lithographer or gravure printer generally makes his own plates, although the process can be subcontracted. The rest of the stages are the same as in letterpress printing.

All of the stages we have described can be found in separate establishments; on the other hand, many printers have the complete range in a single establishment, especially those who specialize in particular end-products, like book manufacture or law printing. These specialized firms may still use subcontractors for part of their requirements, however, perhaps to provide for peak loads or to obtain a special service requiring unusual capital equipment. At the other extreme, there are "printers" who perform none of these stages. These are the printing brokers, whose capital investment consists of a desk and a telephone. They organize the efforts of a number of subcontractors, sometimes having a different firm do each stage of the work.

Other external economies

The availability of these subcontractors is, in the economist's terms, an external economy for the job printer: external to his firm in the sense that the economy does not depend on the size of the firm but rather on the aggregate needs of the industry. More specifically, the availability of the subcontractors is an economy of concentration, that is, an external economy which is dependent on the clustering of many of the industry's firms *in a particular location.*[10] The subcontractors are perhaps the clearest example of concentration economies in printing, but there are a number of other significant and interesting ones.

One of the principal concentration economies available to printing firms in New York or any large printing city is that of external storage for materials. New York printing firms, being located close to suppliers, can rely on them for quick delivery and therefore do not have to hold such large inventories as they otherwise would. This is a saving both on space costs and the carrying costs of the inventory. A survey of printing firms in New York City and in the rest of the United States showed that New York City firms kept 9.7 per cent of their assets (excluding real estate) in the form of inventories, while firms outside of New York kept 19 per cent of their assets in this form.[11]

One New York printing firm has indicated that this was a substantial factor in its decision to stay in lower Manhattan. On considering a move to Long Island, it found that "our increased distance from such sources as paper houses, ink houses, binders, typesetters, etc., would vastly increase the amount of production planning and would necessitate a large inventory of all supplies." [12]

The inventory point is especially applicable to the sort of printing which is concentrated in the city: small operations with unstandardized output, rather than book or periodical printers, for instance. Some idea of the possible diversity of inputs for the printing firm can be gained from the range of the types of paper which these small general printers might be called on to use. In the field of "book paper," which is used for most printing (excluding newspapers and stationery), one study found that there were 40 product classes, 12 grades, 33 finishes, 9 sizes and trims, 8 colors, and 19 types of packing.[13] Obviously a great many combinations of these have never been made or thought of, since the possible total is something more than 170 million, but at least we can assume that a general printer needs access to several thousand varieties of paper. Keeping an inventory of so many kinds is an obvious impossibility for the small printer with his fluctuating demand, but when we consider the demand of New York's printers as a whole, served by the paper supply houses, variations in demand are nowhere near as serious.

Another significant economy for the small and medium-sized firm is the ready availability of repair and maintenance services for printing presses and other equipment. Over 75 per cent of the nation's repair work on printing presses is done by firms located in New York City.[14] Large establishments have a steady enough need to be able to support their own staffs, but small shops have to rely on outside specialists. The cost of a machine breakdown involves not only the cost of repairs, but also the cost of down-time, since the machine is not earning any money. Quick access to specialists reduces the down-time.

The small and medium-size firms in New York are also benefitted by the New York Employing Printers Association, a trade association handling many functions which, in large establishments, would be incorporated in their own operations. The association has a staff of around fifty persons, including accountants, engineering specialists, and labor specialists. They prepare balance sheets, operating statements, and budgets, install cost systems, suggest plant layouts, make appraisals, provide advice on materials, processes, and operations. They negotiate and administer labor contracts and grievances, advise on taxation, insurance, credit and collections, employment and training. In sum, they provide the small printer with fractional use of many services which would not otherwise be available to him.

Finally, with respect to labor supply, New York and other large printing centers have a substantial advantage which is not shared by printers in smaller locations like the one-company towns which are prevalent in book manufacture and in some periodical printing. Isolated printers must deal unaided with the problem of labor-force maintenance. They must provide stability in employment for their workers. A small-town printer cannot hire a pressman for two or three days to help with sudden peak loads of demand, then lay him off and expect to be able to hire him again to meet another peak situation in a few weeks.

Large cities, however, have a printing labor supply which is used by many firms. In addition there are, in New York or Chicago, institutional arrangements which implement the smooth transition of workers from job to job. The various unions maintain out-of-work rooms where workers can register for openings, and the New York Employing Printers Association also maintains an employment bureau at which workers can register their availability and employers indicate their vacancies. The employer can acquire workers for help on a particular job and lay them off when the job is over. The worker can usually find another job within a day. Thus random fluctuations in labor requirements can be met by drawing on the pool of skilled labor through the institutions maintained by unions and employer associations. The size of the labor market itself and the institutional arrangements for transferring labor from one employer to another allow any metropolitan printer to enjoy practically the same economies in this respect that a large employer would have.

The importance of this point is indicated by the case of one book manufacturer who has a printing and binding plant in New York City and another in a small town within the New York Metropolitan Region, about fifty miles from the city. The character of the work differs quite radically in the two establishments. The New York City plant does the work which comes in small and medium-size batches, principally novels and popular nonfiction. The suburban plant concentrates on material for which the press run is considerably longer—for example, subscription books such as encyclopedias. The management states that the principal factor behind this division of work is the differing labor supply situation in the two locations. The suburban establishment is assigned all the work which is of such a nature that the management can plan a smooth and continuous production flow, without marked peaks and valleys, and thus stabilize labor requirements.

❡ Size of establishments within the region

As we pointed out earlier, truck transport permits shipment of printed matter to New York City from locations in the outlying counties of the Metropolitan Region and overnight shipments from cities with lower cost structures several hundred miles away. The complex of subcontractors and other concentration economies, however, makes it difficult for a small printer serving the New York market to locate anywhere else than in the center of the market, even when he is not drawn there by considerations of deadlines and consultation with his customers. In some other place he would be too far from his suppliers to use them for storage or to have ready access to the wide variety of inputs he needs, too far from a labor pool which he could draw on easily for peak loads, too far from his subcontractors to be able to use them routinely.

As individual printing jobs get larger—say the difference between a few thousand copies and a hundred thousand copies—many of the difficulties which we have mentioned fade into insignificance. A printer handling mostly long-run material can economically conduct more stages of the printing operations under one roof—in particular, binding, addressing, and mailing. In other words he can have a more integrated establishment. With longer-run material, too, he can plan his production schedule better, so that he has less need for the flexibility of the urban labor supply and the storage function of paper and ink merchants. Finally, as total time consumed by the printing and binding operations rises, the time involved in consultation (design conferences, checking proof) shrinks as a percentage of total time for the printing job, even if greater distance between the printer and the consumer makes each consultation more difficult.

We might amplify this last point a bit. If the printer is located in a town which we will assume to be in the Metropolitan Region forty miles from New York City, the time consumed in having design conferences and shipping proofs back and forth might be perhaps four times as much as if the printer were located in Manhattan. But the printing and binding time goes up roughly in proportion to the length of the run, while consultation and proofreading times increase only a limited amount. Consequently, the time loss involved in printing one hundred thousand copies in the outlying town instead of New York is much smaller as a percentage of total time than is the time loss for printing a run of five thousand copies there.

As a consequence of all these factors, we can see that the sort of firms which locate in the outlying counties of the Region ought to be on the whole larger, integrated establishments, although we would of course expect to find also in those counties some small printers serving local demands.

We can confirm this view by looking at the difference in the size of establishments between New York City and the other counties in the New York Metropolitan Region taken as a whole—a group which we will call the ring. Table 10 gives figures on average size of establishment for a number of branches of the printing industry, as classified by the Bureau of the Census. The whole first group shows smaller establishments in the city than in the ring. This group includes commercial printing and lithographing, the industries which include the job printers and include also a number of

Table 10 Average Number of Employees per Printing Establishment in New York City and the Rest of the New York Metropolitan Region, 1956

Industry	New York City	Rest of NYMR
Printers		
Book printing	84	98
Commercial printing	18	30
Lithographing	21	37
Subcontractors		
Bookbinding	34	30
Miscellaneous bookbinding	25	18
Typesetting	23	16
Engraving and platemaking	26	16
Photoengraving	50	20
Electrotyping and stereotyping	45	14

Source: Computed from data obtained from the labor departments of New York State, New Jersey, and Connecticut. Connecticut figures are for March 1956, while the others are for September 1956. Industries are classified here according to the Standard Industrial Classification of the federal government (see Appendix C).

printers who specialize in various end-products. Establishments primarily engaged in printing books, though they use the same processes used in commercial printing and lithographing, are separately classified.

The second group, on the other hand, has smaller establishments in the ring than in the city. These industries are all composed of subcontractors for printers. They have a very small amount of employment in the ring to begin with, and the size of establishment in these industries in the ring is essentially dependent on the amount of demand for their product from printers located in the ring. The difference in establishment size for subcontracting industries where it is quite marked between ring and city leads us to suspect that the establishments in the ring in these industries do not have sufficient demand to permit them to operate at their optimal scale, although the cost of the operation they carry out is still lower than it would be if done by the general printers themselves.

Concentration in the City

We can gain another perspective on the difference between the function of the city and the ring by looking at the degree to which various branches of the printing industry are concentrated in the city. In the New York Metropolitan Region, about 50 per cent of all manufacturing employment is located in New York City itself. All the branches of the printing industry are much more centralized in the city than that. But there are differences from branch to branch which reflect differences in their functions in the city.

Table 11 shows three groups of the branches of the printing industry. The first group includes those which function exclusively as subcontractors for other printers or for publishers who have their actual printing done outside of the Region altogether. These industries have almost insignificant amounts of employment located outside the city. The next group, lithographing and commercial printing, is less concentrated in the city because these categories include many firms which specialize in various end products. They can therefore integrate beyond the level permitted to the general

Table 11 Employment Concentration in Branches of Printing in the New York Metropolitan Region, 1956

Industry	Ratio, New York City to NYMR
Subcontractors for printers	
Miscellaneous bookbinding	96.2%
Electrotyping and stereotyping	95.6
Typesetting	92.6
Book and pamphlet binding	90.5
Photoengraving	84.3
General printing	
Lithographing	73.3
Commercial printing	66.4
Book printing	62.8

Source: Computed from data obtained from the labor departments of New York State, New Jersey, and Connecticut. Connecticut figures are for March 1956, while the others are for September 1956.

printer, and are consequently less tied to the city. There are also places in the ring which have their own printing concentrations to serve local demand—Newark, for example. These two industries, then, are less concentrated in the city than the subcontractors. Finally, we have one example of a group specializing in a particular end product, the book printers; they show the lowest concentration in the city of any branch of the printing industry. These establishments do a sort of work less dominated by deadlines than that of general printers. They are also much more highly integrated. Both of these factors lessen the pull of the city.

In summary, then, we have seen that the unstandardized nature of the work done by general printers, together with the short deadlines involved in a good part of the work, has tended to centralize printing close to its customers. Job printing is only one step removed from the publishing and printing of newspapers in the degree of its closeness to the market. But periodicals and books, which are next in line for discussion, display an entirely different locational pattern.

3

Periodicals and Books

PERIODICALS

The periodicals industry[1] includes a wide variety of firms, from integrated industrial giants which own their own forests, paper mills, printing establishments, and distribution facilities, and which issue magazines whose circulations number in the millions, down to tiny one-man-plus-secretary establishments, whose stock in trade consists principally of a typewriter and access to some specialized group of readers such as breeders of fruit-flies. In fact, over half of all periodical establishments have fewer than five employees.

WHERE MAGAZINES ARE PUBLISHED

Table 12 gives a picture of the distribution of the Census periodicals industry, which covers the *publishing* of periodicals and part of the printing—that part which is done by the few periodical firms that still carry on publishing and printing in the same establishment. By far the principal concentration of the industry is in the New York Metropolitan Region, which in this table is represented by the somewhat smaller Standard Metropolitan Area, which had about 27,000 employees in 1954. The area's employment in periodicals, in turn, is concentrated heavily in Manhattan; in fact 93 per cent of it was in Manhattan in 1947,[2] and the situation has not changed materially since then. Outside Manhattan, the only important cluster in the area is in the town of New Castle in Westchester County, where the *Reader's Digest* employs over 2,000 people. The New York Metropolitan Region, as defined for purposes of this study, includes also the Connecticut county of Fairfield (not a part of the Standard Metropolitan Area), which has something under 1,000 employees in periodicals. That portion of New York State outside the metropolitan area has only a few hundred employees in the industry; and the whole of New Jersey, both the

Table 12 Principal Concentrations of the Periodicals Industry,[a] 1939, 1947, and 1954

	1939		1947		1954	
	Employment	Percentage	Employment	Percentage	Employment	Percentage
United States total	61,972	100.0	68,957	100.0	62,428	100.0
New York area [b]	19,176	30.9	25,429	36.9	26,771	42.9
Philadelphia area [c]	5,117	8.3	7,911	11.5	6,776	10.9
Chicago area [c]	6,831	11.0	6,639	9.6	5,983	9.6
Ohio	6,917	11.2	8,629	12.6	5,447	8.7
Iowa	1,423	2.3	1,609	2.3	2,480	4.0
Dist. of Columbia	2,624	4.2	2,405	3.2	2,280	3.6
Missouri	1,726	2.8	1,768	2.6	1,697	2.7
California	2,238	3.6	1,679	2.4	1,547	2.5
Connecticut	d	d	2,520	3.5	1,086	1.7
Rest of nation	....	....	10,368	16.8	11,361	18.2

[a] Excludes many small periodicals ("house organs," for example) because they are attached to other organizations and no separate accounting records for them are available. Includes printing only when done by the periodical firm itself, in the same establishment with publishing. See Appendix C for fuller description of industry.

[b] For 1947 and 1954, the Standard Metropolitan Area (SMA), and for 1939 the Industrial Area, defined in Table 3. The SMA is somewhat larger, but for present purposes there is little difference, since periodicals publishing is primarily located toward the center of the area.

[c] Standard Metropolitan Area.

[d] Not disclosed.

Sources:

For 1939: *1939 Census of Manufactures,* Vol. III, *Statistics for States* (Washington, 1942), Tables 8 and 9 in the section covering each appropriate state.

For 1947 and 1954: *1954 Census of Manufactures,* Vol. III, *Area Statistics* (Washington, 1957), Tables 2 and 5 in the section covering each appropriate state.

parts in the metropolitan area and those outside it, had only 522 employees in the industry in 1954.

Among the other concentrations listed in Table 12, Iowa means principally Des Moines, the home of the Meredith Publishing Company (*Better Homes and Gardens*), and Ohio stands for Cincinnati, Columbus, Cleveland, and Dayton, for which separate figures are incomplete.

Table 12 also clearly shows (for recent years) the trend towards increasing centralization of the industry in the New York area. In 1939 the twelve-county Industrial Area had 30.9 per cent of the nation's employment in periodicals, and by 1954 the Standard Metropolitan Area had 42.9. (Since most of this employment is concentrated in Manhattan, the fact that the Standard Metropolitan Area is bigger than the Industrial Area makes little difference.) From 1939 to 1947 almost all the growth in the national total for the industry was accounted for by New York. From 1947 to 1954, the change was even more striking: the rest of the nation suffered an absolute decline, while New York was practically the only area which continued to grow.

Much of the industry outside New York consists of either small, scattered publications of the type which do not have urgent need for the economies and other advantages of a New York location, or printing employees of periodical publishers that still do their own printing. In the periodicals industry nationally, 25 per cent of the employment is "production workers," most of whom are printing employees. The corresponding figure for the New York Standard Metropolitan Area is 16 per cent; [3] so New York's share of the nation's nonproduction workers in the periodicals industry runs higher than its share of the whole industry—indeed about 48 per cent.

New York's concentration is even more impressive if we look at magazines of large circulation. Table 13 shows that 70 per cent of the magazines with circulations above 200,000 are published from the New York Metropolitan Region—86 out of 121.[4] A number of the magazines which are represented as falling outside of the New

York Metropolitan Region are magazines which our later analysis will indicate have substantially less reason for locating in New York than some others. Principal examples are a number of large religious monthlies and weeklies, and a number of specialized magazines (like *Motor Trend*) and fraternal journals (*Rotarian*).

The pull of the area is seen even more clearly when we consider new magazines. The vast majority start in New York, many as

Table 13 Concentration of Publishing Offices of Large-Circulation Consumer Magazines in the New York Metropolitan Region, 1957

Circulation	Number published in U.S.	Number published in NYMR
Over 10 million	1	1
5 to 10 million	2	1
3 to 5 million	10	7
1 to 3 million	26	21
500,000 to 1 million	31	19
200,000 to 500,000	51	37
Total over 200,000	121	86

Source: Audit Bureau of Circulation, publishers' statements to October 1, 1957.

new enterprises of established concerns: *Sports Illustrated* within the Luce empire or *Purchasing Week* by McGraw-Hill, for instance. Others are formed by staff members of existing magazines who break out to start their own publications.

A number of magazines of course have been started outside New York, but the trend, almost without significant exceptions, is for them to move to New York when they begin to feel growing pains elsewhere. Good examples are *Esquire* and *Coronet* (from Chicago), *Woman's Home Companion* (from Springfield, Ohio), *Time* (which tried Cleveland for a period in its early history and then returned to New York), the Cowles publications (from Des Moines), and the Fawcett publications, which moved from Minneapolis to Greenwich, Connecticut, and later into Manhattan itself. There are few cases of any significance where a magazine has started in New York and moved elsewhere.

What accounts for the heavy concentration of periodicals in New York? In the first chapter we suggested that the primary reason was the cluster of talent in the area, which in turn is connected with New York's early start as one of the larger publishing centers and its status as a first city. As the publishing center developed in New York, many economies of concentration came into being, and these further enhanced New York's desirability as a location. Rather than trying to untangle the whole historical process, however, let us look at New York as a possible location from the perspective which new publishers might have.

EXTERNAL ECONOMIES

The New York area is an ideal location for external economies. What this means for magazines in particular, aside from the economies usually available to employers in a city location, is that the New York area has many agencies and groups offering specialized services of particular utility to magazine publishers. There are the research facilities of New York: the great libraries like New York Public Library and the Columbia University Library and the nation's largest collection of libraries specializing in particular fields.[5] Then come subcontractors who are specialized for service to magazines and to other communications industries. For example, there are agencies which maintain comprehensive files of pictures of all conceivable people, scenes, and subjects, available on short notice for use in magazines. There are agencies which handle with electronic equipment the complex problems of circulation fulfillment (billing, labeling, accounting, and mailing) for small periodicals. There is a group of agencies which maintain mailing lists of people with particular interests; publishers are finding such lists increasingly useful for publicizing articles on specialized subjects or advertising the magazine itself to particular groups. There are free-lance specialists in art work, photography, layout, and translations.

There are literary agents and cartoon agencies. Some of these facilities can be duplicated elsewhere, but most cannot. The inter-knit complex taken as a whole is a unique feature of the New York area.

These services are performed outside most magazine offices because they can be done in that way at lower cost than when performed by the magazines individually. Even the largest magazines use some of the services. In certain fields, the economies, in terms of massing of resources, continue well beyond the scale which any publisher could afford—in libraries and photo agencies, for instance. Smaller publishers benefit from the whole range of services. The broader a magazine's subject-matter coverage, the greater is the need for these services and the closer to them it must be. A specialized magazine—a trade journal or a hobbies magazine—may use such services only occasionally, but a general-interest magazine will have continued need of them.

The literary labor force

Perhaps more important than these external economies is New York's pool of specialized labor. New York and its surrounding suburbs and exurbs have the largest concentration of writers, editors, and artists in the United States. The writers fall into two groups: those who serve on the staffs of magazines, and free-lance writers. The concentration of staff writers is obvious from the concentration of the industry in the New York Metropolitan Region. The free lances are also located in the Region. The Society of Magazine Writers, which includes most persons who make their living principally by writing free-lance material for the magazines, estimates that 75 per cent of its membership lives in the New York Metropolitan Region.[6] Besides, New York's big newspapers provide a training school from which there is a continual flow of talent into the magazine field.

The availability of New York's pool of labor adds significantly to the flexibility of magazine publishers and allows them to expand and contract operations without undue difficulty. The attraction of the labor pool is doubly strong for magazines with large staff requirements for writers. Location outside of the Metropolitan Region would make it difficult to fill staff positions as openings come up, and the problem is especially acute for magazines just starting out.

Starting a magazine does not merely mean hiring a staff, producing a magazine, and then selling it. Preparation of a new periodical is always a gamble, and many of them never get beyond the dummy stage, or beyond a few issues if they do see the light of the newsstands. Therefore the problem faced by those launching a new magazine is this: they must be able to assemble at least a skeleton staff, without necessarily making any promises about long-term employment opportunities. Obviously this is hard to do if one is working from Portland, Maine, or the middle of Kansas, however attractive these locations may be in other respects. Frequently these trial flights have to be staffed by writers who are specialists in the particular field concerned, perhaps politics or music or men's apparel. A new publication locating in New York can acquire staff, specialized and otherwise, much more easily than elsewhere.

Many publishers also experiment with single-issue magazines, called "one-shots," in order to capitalize on subjects which have caught the public's fancy: do-it-yourself and photography, James Dean or Elvis, hot-rods and rockets. It is estimated that in 1956, publishers issued about 500 one-shots selling over 14 million copies.[7] Some publishers (Dell, Fawcett, Macfadden) issue them occasionally along with their regular publications, and there are several publishers like Skyline Features Syndicate and Great American Publications which specialize in the one-shots. The primary factor is speed: each publisher tries to get his one-shot on the stands before someone else's one-shot fires first. This kind of production, like that of the regular magazines, requires a large supply—and more importantly, a varied one—of skilled creative personnel especially on a free-lance basis.

From the viewpoint of the magazine employee, there are excellent reasons to come to New York—or to stay there despite offers from elsewhere. One is the broad range of alternative opportunities for employment when a new job becomes either a pleasant thought or a necessity. These opportunities are not only on the magazines, but also in related industries: for example, book publishing, newspapers, advertising, foundations, national organizations and trade associations with New York headquarters, and public relations departments of large corporations.

Another characteristic of New York is the sort of people it attracts: intellectuals and imitations of intellectuals. New York serves as a magnet for the talented, to some degree independently of the specific availability of jobs. Many find to their surprise that they end up in nine-to-five writing jobs. The availability of many writers in the New York area (and not elsewhere) results from the fact that they are in a way "externally conditioned labor," in the phrase of Abbot P. Usher, one of the early specialists in the economics of location.[8] They may be in New York because of the atmosphere—the Village, the museums, the excitement of Mecca, the surplus of unmarried females eighteen to twenty-five, the air of promise for the young intellectual with his wits freshly polished by expensive universities.

Time discovered this situation when it moved operations from New York to Cleveland in 1925, attracted by lower printing costs. Many of the younger staff members found life less exciting in the provinces, and *Time* missed the continual supply of bright young men which New York had to offer. The magazine returned to Manhattan in 1927.

✓ Communication

The combination of New York's specialized labor pool and its external economies—subcontractors in particular, specialized for service to the magazine industry—accounts for the concentration of periodicals in the New York Metropolitan Region. But the question still remains: why Manhattan rather than elsewhere in the Region? The availability of subcontractors and the necessary labor supply would not seem to rule out Englewood or Greenwich or places even further in the periphery. After all, many of the writers, editors, and artists now prefer to live in Exurbia, the wider-open spaces beyond Suburbia, and a location outside the city might offer them distinct commuting advantages.

The answer is to be found principally in the cost of personal communications and the time it takes to carry them out. We have already discussed in the historical chapter our belief that the chief factor causing the concentration in New York of the periodicals industry in the late nineteenth and early twentieth centuries was the necessity of minimizing costs of communication among publisher, author, and artist. The question then arises—why should this still be a factor in the middle of the twentieth century, when telephone, telegraph, special delivery mail, the airplane, and even closed-circuit TV make swift communication possible from any part of the country to any other? The reason is partly that these forms of communication are all costly, but the answer lies more basically in the nature of communication within the periodicals industry.

Much of what goes on in magazine publishing depends very heavily on informal contacts of a wide variety: trade gossip exchanged over lunch (in a business where gossip *is* the trade), office meetings with writers and artists, tips on the grapevine. These contacts provide leads for potential materials to use in the magazine and advance dope on what the competition is planning. Many of these contacts have no substitutes elsewhere, even through the virtues of modern electronics. It is a question not of relaying specific messages, which one might arrange to come over the phone at two-fifteen every day, but a whole atmosphere of informal, personal, and immediate communication, outside of office hours as well as within. If a publisher is not in Manhattan, his staff misses the fruits of these contacts.

The communication question, however, goes beyond the problem of contact with other people working within the magazine industry, important though that phase of the operation is. There is also the question of communication with New York's variety of news sources for the widest variety of subjects. New York, as the nation's first city, plays a unique role in this respect. It is a prime generator of news and information (for periodicals as well as for newspapers) in practically all fields of general interest: not only political news, but financial, cultural, and so on. The United Nations makes it the focus of much international political news. Wall Street is the nation's financial center; Broadway is the nation's theater. The Garment Center is America's fashion dictator. Foundations have their headquarters in New York. The principal commercial art galleries are there, as well as many of the nation's most important endowed galleries. Washington is another great news center—the country's greatest source of political and governmental news. But Washington's news sources are profusely covered by the wire services, whereas much of New York's broad spectrum of activity is not.

This analysis of the communication factor applies with particular force to the most competitive sector of the periodicals industry: general large-circulation periodicals like *Life, Look, Time,* and *Newsweek,* which aim at keeping up with a wide variety of subjects. This competition can be translated principally in terms of speed. These periodicals are in competition to "do" Ingrid Bergman not only best but also first when she pops up in the news: the most facts, the best photographs, the latest gossip. Each strives to keep a week ahead of its rivals, and to maintain the quality and timeliness of its reporting on a wide field of topics.

There are of course exceptions to the general rule that magazines must be in New York. There are, first of all, a few large-circulation general-interest magazines which are not centered in New York. There is the Curtis group, published in Philadelphia: *Saturday Evening Post, Holiday, Ladies' Home Journal,* and *Jack and Jill.* The location of the Curtis empire in Philadelphia seems more the result of historical accident than anything else. Once an enterprise of that size has grown up at a given location, the barriers to a change are considerable: for example, the attachment of trained employees to the location, and the ownership of specialized physical facilities.[9] Even so, the high executives of the Curtis magazines pay frequent visits to New York.

The *Reader's Digest,* located within the New York Metropolitan

Region about 35 miles from New York City, is a case of historical accident combined with less exigent editorial requirements. At the start, the *Reader's Digest* was simply a magazine which trimmed the excess fat from articles appearing in other magazines. The location in Westchester was a matter of personal preference for the editors, and, as the magazine grew, it had little need for Manhattan's communication economies, since its operations involved mainly people, a stack of current magazines, and an ample supply of blue pencils. There was no need, then, for a location in the center of the metropolis, and the Westchester location still allowed the *Digest* to draw on New York's pool of literary labor.

Competitive, general-interest consumer magazines form only a part of the total periodicals picture, although certainly the largest part. The specialized magazines that make up the rest of the magazine industry have much less need of a location in the New York Metropolitan Region. Specialized magazines of various sorts are more likely to locate outside of the New York area, for a number of reasons. Because they do specialize, they can be viewed, in an economist's terms, as producers who have a degree of monopoly and monopsony power which insulates them from competitive pressures and makes the time factor of much less importance. Many of these specialized magazines are indeed located in or near Manhattan, because the same external economies are available to them as to the general-interest magazines. But they do not have to be in the center of the sources of all types of information; they are drawn to the specialized centers which particularly interest them, and which have complexes of informal communications concerned with their own fields.

Many of these specialized magazines are published as adjuncts of other programs or organizations, such as research projects, universities, and national headquarters of various kinds. In these cases, the basic locational problem is not where to put the magazine, but where to locate the parent organization, with the result that the magazines are scattered all over the country. Trade journals often locate near the regional centers of their industries. *Electrical West,* a journal for the Pacific Basin electrical industry, is published in San Francisco, and a number of printing trade journals are published in the Midwest. But even for trade journals which have regional news sources and regional subscribers, New York has attractions. McGraw-Hill Publishing Co., the largest publisher of trade journals, has central offices for all but a few of its trade periodicals in the McGraw-Hill building in Manhattan, where the staff of each can take advantage of common overhead facilities such as a photography department, production control, advertising solicitation, and circulation facilities. McGraw-Hill maintains field editorial offices, however, in regional centers where necessary. *Petroleum Week*, for instance, has field offices in Tulsa, Houston, and Dallas.

Our discussion up to this point has concerned principally the editorial and art departments of periodicals. But the same considerations apply to their advertising offices—and perhaps with greater force, since many periodicals which locate their editorial offices elsewhere nevertheless are constrained to maintain advertising offices in New York City in order to be able to compete effectively for the consideration of the advertising agencies. The *Reader's Digest,* although located in Westchester, opened advertising offices in Manhattan when it began to take advertising a few years ago. *Electrical West* has an advertising office in New York. The Curtis publications, big as they are, maintain branch advertising offices in New York.

As shown in Table 14, the advertising agencies in the New York Standard Metropolitan Area take in almost half of the receipts of advertising agencies in the United States—a concentration approached by no other city. Advertising provided in 1954 almost 62 per cent of the total income of the magazine industry as a whole, and for many periodicals the fraction is a good deal higher, since the average includes many periodicals which do not accept advertising or carry very little of it.[10]

Consultation with advertising agencies is not easily carried out at a distance. Much of the contact between agencies and magazines

Table 14 Principal Metropolitan Areas in Receipts of Advertising Agencies, 1954

		Receipts		Employment	
	Establishments	Thousands of dollars	Percentage	Number	Percentage
United States total	3,267	3,165,724	100.0	45,607	100.0
New York	769	1,469,814	46.4	19,696	43.1
Chicago	320	494,280	15.6	5,861	12.9
Detroit	94	202,432	6.4	1,913	4.2
Los Angeles	227	101,323	3.2	1,718	3.8
San Francisco-Oakland	105	89,888	2.8	1,105	2.4
Philadelphia	116	84,718	2.7	1,952	4.3
Rest of nation	1,636	723,269	22.9	13,362	29.3

Note: Standard Metropolitan Areas as used by the Census Bureau. Advertising agencies covered in this table are, in the technical language of the Census Bureau, "Advertising Agencies with Payroll," thus excluding one-man offices.

Source: *1954 Census of Business,* Vol. V, *Selected Service Trades—Summary Statistics* (Washington, 1956), Table 9A, p. 9–2.

has to be of a highly personal, face-to-face nature, of the sort we have already discussed with respect to the editorial departments. The magazine's product, in this case advertising space, has to be kept constantly before the eyes of its buyers.

In addition, layouts must be planned, type set, engravings manufactured, plates made, proofs checked. The whole process requires close consultation among ad agency, magazine, and various branches of the printing service trades. A great deal of this work has to be done on a tight schedule. Consequently any magazine whose advertising department is not at the center of the industry suffers from a competitive disadvantage in terms of communication.

✓ Periodical printing

The printing of periodicals has behaved very differently from the publishing of periodicals. Although the printing was originally located close to the publishing, making New York the periodical-printing center of the country as well as the publishing capital, the course of the twentieth century brought certain developments, which we have already mentioned in the first chapter of this study, permitting the separation of printing from publishing in order to take advantage of lower labor cost and the lower transport cost of serving a national market from the Midwest. The 1918 zoned rates for periodicals (to be discussed further in the next chapter), combined with labor unrest in New York and decreasing unionism elsewhere, were the specific impetus. Improved means of communication were the permissive factor.

Although some periodical printing had left the city before 1920,[11] it was the early 1920's that saw the beginning of a substantial change. A series of moves out of the city began in 1922. If we list all the large printing firms that existed in New York City at any time between 1922 and 1940, we find that the twelve largest of these—all but two of them primarily periodical printers—had left the city or folded by the end of the period. Four went bankrupt as a result of the migration of periodical printing westwards; two moved to the Midwest themselves; five others moved to Pennsylvania or to Albany, New York; and the remaining one closed down to reappear as two establishments elsewhere, one in Dunellen, New Jersey (in the New York Metropolitan Region), the other in Chicago.[12]

The disappearance of these large concerns from New York City tells only part of what has ultimately become of periodical printing. Although the era of bankruptcies and large moves was over by World War II, periodical printing continued to move to the Midwest. In general, printing firms do not move, but printing work does.

Since all printing of periodicals is not grouped together in a separate Census industry, employment figures are unavailable. But when we look at commercial printing receipts which are derived from the printing of periodicals, we see that by 1954 almost 42 per cent of them went to printers in the East North Central region. (See Table 19, next chapter.) These firms were principally in Ohio and Illinois. Only 30 per cent went to commercial printers in New

York State and New Jersey. More than 25 per cent of Illinois' commercial printing is printing of periodicals, while only 10 per cent of New York State's is periodicals, a fraction which is a shade lower than the national average.[13]

The migration from New York can be indicated dramatically by the change in New York's printing equipment. Long-run magazines are printed on high-speed rotary web presses. In 1920, aside from newspaper presses, there were 177 of these rotary web presses operating in New York. By 1929 the figure had fallen to 60, and by 1940 there were less than 15 of them in the city.[14] Not a single large national periodical is printed in New York City at the present time, whereas in 1920 almost all were.

PRINTING BY REMOTE CONTROL

While the publishers remain in New York, the development of cheaper and faster transportation has made possible the location in the Midwest of the manufacturing stage of magazine publishing. Air mail compressed distances, and the teletype compressed them even further. And the teletypesetter, whereby a person at a keyboard in one city can set type in another, has become increasingly important in the last two decades.

Time provides a good example of printing by remote control. The domestic edition is printed at four contract printing plants—in Chicago, Los Angeles, Philadelphia, and Washington.[15] Editorial matter is sent by teletypesetter to the plant in Chicago; the pages are made up there, after which mats or plates are flown to the other cities. Films of the pages also go by air to Paris, Tokyo, and Havana for the printing of international editions.

Newsweek sends editorial matter by teletypesetter from New York to its contract printing plant in Dayton, Ohio. McGraw-Hill's printing of its many technical and trade journals is done in printing plants scattered in a number of different localities, but the editorial offices maintain close contact by teletype. Still other publishers in New York fly negatives, proofs, or mats (from which stereotype or electrotype plates can be made) to printing plants in the Midwest.

The printing proper is not tied to a city location, although a number of the Midwestern printers are in cities. A fairly large percentage of periodical printing is done in the outlying counties of metropolitan areas—or outside of them altogether. This is so because of a key characteristic of periodical printing, standardization, which differentiates it sharply from the sort of work which remains in cities. Although each issue of a magazine is a separate commodity with its own problems and idiosyncracies of design and production, many elements of a magazine present no design and decision problems, week-to-week or month-to-month. Format remains the same (a matter of some importance), paper requirements are standardized and predictable, type faces are standardized, and other production specifications are relatively invariant. Therefore there is not the same need for the variety of subcontractors which we find with a general printing firm producing a highly unstandardized product. The size of the operation also works towards the same end. Periodicals are printed in long runs, thus permitting organization on a much larger scale. The printer is able to perform at an economical scale many of the operations which a smaller printer must contract out. Also because of the long runs, a printer specializing in periodicals and similar work can operate with a much smoother production schedule and fewer peak-load problems than a general printer, a situation which reduces the need for a large community pool of labor.

The unstandardized stages of printing which demand consultation, checking, and examination, however, often remain in New York: engravings, which have to be checked and rechecked; the typesetting and layout of advertisements, involving advertising agencies, publishers, and sometimes two or three varieties of printers; and the typesetting of text material, which has to be proofread and worked into a layout involving pictures and drawings. *Look*, for instance, has all its typesetting done twice. It is done initially in New York. Proofs are made and worked into page layouts, which are then airmailed to Chicago, where typesetters at the printing plant set an exact facsimile of the type as set in New York.

This separation of earlier and later stages helps to account for the fact that the New York area has larger shares of the nation's typesetting and photoengraving than of commercial printing and lithographing. The percentages of national employment in the Standard Metropolitan Area are as follows:[16]

Commercial printing	15.9
Lithographing	17.6
Typesetting	32.3
Photoengraving	26.9

The final department of magazine production which deserves mention is subscription fulfillment. This routine operation, involving unskilled clerical labor, is tending to move out of New York City and to some extent out of the Region altogether. The stages of billing, accounting, and preparation of address plates can take place practically anywhere, and publishers have located these stages in various cities in order to find low-cost labor. Des Moines, Iowa; Boulder, Colorado; and Philadelphia are examples. The final addressing, however, is done of necessity at the printing plant.

Periodical printing and publishing, then, illustrate an interesting tendency of a number of industries originally located in New York. It has been locationally split, allowing each stage to seek its own least-cost location. This separation has largely been the consequence of improvements in transport and the means of communication.

BOOK PUBLISHING AND PRINTING

As in periodical publishing, New York is today the overwhelming leader in book publishing in the United States.[17] At the latest Census in 1954, the New York Standard Metropolitan Area had 324 book publishing establishments, employing more than 15,000 workers—almost as many as the rest of the country put together. No other center even approaches it in importance. Table 15 shows the leading states in book publishing. Almost all the employment shown for New York State is within the New York Metropolitan Region. In terms of numbers of copies published, the concentration of American book publishing in the New York Region is even higher, above half of the national total.

New York's dominance represents the continuation of the centralization which went on in the nineteenth century. The earliest

Table 15 Concentration of Book Publishing, by Leading States, 1954

	Establishments	Employment	
		Number	Percentage
United States total	814	34,704	100.0
New York State	333	15,929	45.9
Illinois	88	5,887	17.0
Wisconsin	15	2,080	6.0
Pennsylvania	39	1,673	4.8
Minnesota	23	1,526	4.4
Massachusetts	39	1,412	4.0
Rest of nation	277	6,197	17.9

Note: See note to Table 16.

Source: *1954 Census of Manufactures,* Vol. II, Part 1 (Washington, 1957), Table 2, p. 27A–6.

available figures for book publishing given separately from job printing are for 1919,[18] when the New York City figures covered almost all of the book publishing in what is now the New York Standard Metropolitan Area. As Table 16 shows, there has been a substantial increase in the concentration in New York since 1919, at which time New York unquestionably had long been the book publishing capital.

Book publishing can be divided for our purposes into two significant parts: trade books and nontrade books. Trade books, as the name implies, are those generally sold through book stores: fiction and popular nonfiction. Nontrade books include everything else: textbooks and scholarly books, Bibles and other religious books, and reference works.

Since the term "trade books" is necessarily a bit vague, a few more clarifying remarks may help. In the words of William Miller, "trade books" is "the industry's jargon for general literature."[19] Although the name "trade books" originally came from the fact that they were distributed through ordinary booktrade channels, they now are no longer confined to those channels. The book clubs, for instance, now distribute more than a quarter of the output of trade books. The distinction in method of distribution is not important for our purposes, since all we want to do is to differentiate publishers whose output is predominantly of trade books from those whose interest is primarily in other fields.

Table 16 Concentration of Book Publishing, 1919 and 1954

	Establishments	All employment	Production workers
1919			
United States total ...	829	18,557	6,366
New York City	236	5,617	1,269
Ratio, New York City to United States	28.5%	30.3%	19.9%
1954			
United States total ...	814	34,704	8,802
New York Standard Metropolitan Area .	324	15,065	1,853
Ratio, New York SMA to United States	39.9%	43.4%	21.1%

Note: The industry for which statistics are presented here is "book publishing and printing," which, however, includes only book printing done in the same establishment with publishing. The fraction of this industry devoted to printing is relatively small, especially in the New York Area, as indicated by the above fraction of employment representing "production workers." See Appendix C for classification details.

Sources: *Fourteenth Census of the United States* (Washington, 1922), Vol. VIII, *Manufactures, 1919: General Report,* Table 52; Vol. IX, *Manufactures, 1919: Reports for States,* Table 59; *1954 Census of Manufactures,* Vol. II, Part 1 (Washington, 1957), Table 2. Also special tabulation in our Appendix D, Table D-1.

An idea of the importance of trade books can be gained from the Census breakdown of the book-publishing industry's sales for 1954:[20]

Total book sales	$628,551,000
Trade books	220,160,000
Textbooks and workbooks	180,445,000
Encyclopedias and miscellaneous	128,179,000
Technical, scientific, professional	63,635,000
Religious	36,132,000

Trade books, then, form over one-third of the output of the industry, and are more important than any other category. Many firms publish both trade and nontrade books, but the difference between the types is locationally significant. Publishers who are predominantly trade book publishers are much more drawn to New York City than are the various sorts of specialized publishers.

Trade publishing

The principal publishers of trade books are located in New York, with only two exceptions, both in Boston, one being Little, Brown and the other Houghton, Mifflin. Both of them maintain branch offices in New York. There are three other publishers outside of New York City whose output of trade books is substantial: the Bobbs-Merrill Co. of Indianapolis, the World Publishing Co. of Cleveland, and J. B. Lippincott of Philadelphia. Each of these houses has large departments in other publishing fields. The significant point is that all three have their trade book departments in New York City, although their other departments are at their home locations.

Trade book publishers are exceedingly dependent on the economies of communication which go with a New York City location. In book publishing, trade publishers play the same role which the publishers of general-interest consumer magazines do in the periodicals field. They have to be always on the alert for new writers, new subjects, new ways of tickling the public palate. In order to do this they must be in fairly intimate contact with the grapevine of the publishing world. Their market for raw materials is New York. The city has most of the literary agents in the country, and is the center of an area in which an indeterminate but unquestionably high percentage of all the literary people of the country live. In a sense, the market for the final product is also New York. Most of the principal literary magazines of the country are published there, and they help to determine the buying habits of the country at large. Intimate contact between publisher, editors, and reviewers frequently seems to be essential in launching a popular success. Contact with the book clubs and with paperback reprint houses is also important, since much of publishers' income in the last twenty years has depended on the successful negotiation and sale of reprint rights. The whole trade publishing business is one which depends to an extraordinary degree on personal contact and personal acquaintance with the people involved. Contacts of this sort would be difficult to maintain without location in New York.

Because of the importance of informal face-to-face contact in the industry, publishers have clustered in a narrow area of midtown Manhattan. For instance, Prentice-Hall, one of the largest firms, which has recently moved most of its operations to Englewood Cliffs, just across the Hudson River, has nevertheless found it necessary to maintain offices in Manhattan for top people in the trade book department.

Other factors as well draw these publishers to the New York area, factors we have already discussed for periodical publishing. New York's literary labor force is naturally a strong attraction for book publishers as well as periodical publishers. We have already mentioned the concentration of authors in the area. In addition, of course, the publishers must staff their own editorial operations. The principal market for manuscript readers, editors, and artists is in New York as well.

For smaller publishers, the availability of outside contractors for various aspects of publishing is important, too. Editing, translating, art work, layout and design are done by a variety of free-lance people and agencies in New York. There are contract specialists in book warehousing and distribution, and specialists in the design of advertising for book firms and its placement in periodicals and newspapers. A number of the principal book printers are located in New York, and most of the principal ones outside of New York maintain New York offices.

An examination of the sort of book publishing which has remained outside of New York leads us to believe, however, that the most important factor involved in centralizing the book publishing industry has been the necessity of personal communication at low cost.

✔ Specialist publishing

Publishers of nontrade books have considerably more freedom in location than trade publishers. As Hellmut Lehmann-Haupt notes:

> New York was and still is the seat of a number of important special publishers in various fields. However, it is very important to note that there has been no particular concentration of special publishing in New York City. In marked contrast to the trends observed in general publishing, many of the houses devoted to a particular, well-defined field have been founded in various centers throughout the country, where many of them have remained to this day.[21]

Specialist publishers of various sorts (law books, medical books, textbooks, technical books) are dotted all over the map of the country. They have very little dependence on New York's face-to-face contacts and influence. They have no concern with the current literary scene, nor are they dependent on ordinary booktrade channels of distribution and advertising. They are marketing the works of authors who are, for the most part, not professional literary people and not concentrated in New York. The channels of influence of these publishers are largely through salesmen and the mail, rather than through literary periodicals.

A look at the roster of publishers in Boston, Chicago, and Philadelphia—the three minor publishing centers—reveals that the publishers located there are almost without exception specialist publishers of various sorts.[22]

The university presses make up another significant group that is not tied to New York. Although they do produce many trade books, they concentrate on scholarly works and are neither dependent on the Manhattan environment nor able to go there in any case, since they locate in the university communities which they serve.

Balanced against the strong concentration of publishing operations in Manhattan is the fact that a number of the largest publishers have located their warehousing facilities and, in some instances, some of their routine office operations outside of the city. There has been a strong trend in the last five or six years towards the location of these routine activities on the New Jersey side of the Metropolitan Region. Four of the largest publishers have moved parts of their operations to New Jersey: Prentice-Hall to Englewood Cliffs, Oxford University Press to Fairlawn, McGraw-Hill to East Windsor, just outside of the Metropolitan Region, and the Macmillan Co. to Camden, well outside the Region. Prentice-Hall moved everything except its principal trade-book editorial offices, Oxford moved accounting as well as warehousing, and the other two are, for the moment, just moving warehousing, but the possibility of moving other operations is still open.

The McGraw-Hill example illustrates what is basically involved in all of these warehousing cases. Prior to the move, McGraw-Hill's warehousing operations were scattered all over the eastern seaboard. The company employed about twenty bookbinders and stored nearly 2,000,000 volumes (either bound or in sheets) with these binders, calling them into its New York warehouses in small lots as it needed them, and paying relatively high storage charges at the binders' crowded warehouses. On arrival in New York, the books were put in one of three warehouses in Manhattan, and then transshipped to the final warehouse and shipping room at the main McGraw-Hill building at 42nd Street, where space was again short. In order to eliminate the high charges for storage at the binders, and to avoid double and triple handling, the company decided to consolidate its warehousing activities. There was the possibility of concentrating operations in the main building, or building on land adjacent to it owned by the company, but it was felt that a higher return could be obtained for this space if it were used for offices rather than for warehousing.

A further—and perhaps more important—factor working against location in the city was congestion at the New York post office, which often slowed down shipments by days at a time. Therefore it was decided to move operations out of the city. This move would permit one-story construction for smooth materials flow and mechanization of handling, and the incorporation of a post-office unit within the building. Mechanization permitted by the new layout is expected to cut manpower required for shipping operations by two-thirds.

The new building is located just off the New Jersey Turnpike, at the junction with another truck route, permitting the conversion of all shipping to trucks. Shipments going south and west will go directly out without passing through the bottleneck of the New York post office and can be pre-sorted and packed prior to shipment for easy handling when they arrive at a post-office break-up point. Many of the trucks, for instance, will go directly to the Philadelphia post office.

Book printing

Book printing is another stage in the production of books which has been locationally detached from the publishing stage. The printing is generally carried out by printing firms on contract to the publishers, rather than by the publishers themselves. Relatively few publishers do their own printing any more.[23] The separation is geographic as well as organizational. Although more than half of the output of the book publishing industry is published from New York, something less than 30 per cent of it is bound and printed in the New York Metropolitan Region.

The chief factor pushing book printing out of the New York area has been high labor cost—the same as for other varieties of printing. But book printing has one additional feature which frees it from the pull of the Metropolitan Region: it is on the whole much less demanding in terms of time than most other forms of printing. Deadlines in book printing are more often in terms of seasons of the year—the Christmas rush or school opening—rather than a particular week or day. Indeed, for some products in the book industry which have no close competitors, like specialized textbooks or art books, American publishers have had their printing done abroad, where wage levels are substantially lower. The United States copyright law, however, imposes severe restrictions on the importation of books written by Americans and printed abroad.

In any case, most book printing is not quite leisurely enough to permit transatlantic subcontracting, but the phenomenon in these few instances is illustrative of the general situation. Book printing is not usually tied to the source of its copy by tight deadlines or necessity for quick consultation between printer and publisher. Although much consultation is of course involved, the more elastic deadlines make it possible to carry out a great deal of this through the mails. The book printer can also do more advance planning than a job printer can; the inputs are nowhere near as diverse.

Nevertheless, there are similarities between the sort of book printing which has located outside of the Metropolitan Region and the sort of general printing which has moved out. Long-run book printing in particular is relatively little done in the New York area any more—and especially not in the city proper. The printing of cheap paperback books is done almost exclusively outside the Region. These paperbacks have initial press runs of two hundred thousand copies, compared with a few thousand for the average first novel. Book club printings, which are often many times the size of the original printings of the same books, are done mostly outside the Region as well. Where runs are shorter and deadlines more pressing, we find more of the printing done in the New York area. Concentration of book printing in the city proper is lower than for any other branch of printing, however, as Table 11 in the second chapter demonstrates.

Book printing has shown less tendency to migrate to the center of the national market than has periodical printing, even though both products are distributed nationally. As work has left New York, the gainers have been small towns in the eastern part of the country. Such places as Clinton and Norwood, Massachusetts; Scranton and York, Pennsylvania; Binghamton, New York; and Kingsport, Tennessee, have risen to prominence as book printing towns in the last three decades.

Book printing has remained along the Atlantic seaboard to a greater extent than other kinds of printing. In 1954, the Atlantic seaboard states accounted for 54.6 per cent of revenues from the printing and binding of books and pamphlets, but only 43.2 per cent of commercial printing and lithographing taken together.[24] We will examine the reasons for this difference in more detail in the following chapter. The chief reason is the fact that postal rates for books are not distance-sensitive, but flat rates for the entire country, so that location with respect to the ultimate consumers of books is not a cost problem. The opposite is the case with periodicals and catalogues and directories.

The book industry, then, repeats the essential characteristics of the periodicals industry. The publishing stages of both are principally in Manhattan to take advantage of New York as a center of communication and ideas. Routine stages of the industries, however, like warehousing of books and printing of both books and periodicals, are not drawn to the communication center, but are free to seek out lower-cost locations elsewhere. In periodical printing, these locations are determined by the cost of labor and the cost of transporting the final product, taken together. In book printing, labor cost is the principal factor of importance.

4

Printing Costs

The purpose of this chapter is to present a picture of the regional variability of the cost structure for printing, in order to understand better the forces determining the geographical distribution of the nation's printing activity. Throughout the discussion we must hold one thing in mind: the variability of the cost structure can be influential only for certain categories of printing. In much printing work, as we indicated in Chapter 2 of this study, lower costs available at other locations than that of the originating office are not relevant, because of the tight deadlines and the large amount of consultation necessary in the production. The printing simply has to be done close to its originating office, whether this be in New York City, Chicago, or North Adams, Massachusetts. The extreme example is newspaper printing, which is tied close to the publishing office and to the local market which the newspaper serves.

For non-newspaper material, however, as the length of run increases, factors of time and consultation become less influential and geographical variation in cost becomes more important in determining the location of printing activity.

PRINTING COSTS IN GENERAL

The cost structure for printing has three principal elements: materials, labor cost, and transportation. Given the available data, it will be most convenient to deal first with the cost of materials and labor, and then see what difference the cost of transportation makes.

Table 17 shows printing costs for establishments specializing in book printing, in commercial printing (letterpress), and lithographing (offset). These categories include the bulk of the printing work of the country. Newspapers are excluded, as well as bookbinding and the service industries (typesetting, engraving, photoengraving, electrotyping and stereotyping). Almost everything else falls within the three categories presented in the table, including nearly all of the printing of periodicals, catalogues and directories.

The first item listed in the table is materials. Materials are an important locational force in some industries because of their different costs at different locations. In printing, however, the problem is quite simple. Paper is the only important material with respect to weight, and throughout the broad northeastern belt of the country, the delivered price of paper is identical from location to location, in spite of the fact that paper is a heavy material in relation to its value. The belt includes the states bounded by (and including) North Carolina, Tennessee, Missouri, Iowa, and Minnesota. This territory, in which about 80 per cent of the country's printing is done, is Zone No. 1 in a system of zoned delivered prices which is maintained by the manufacturers of book paper. Book paper is used for books, periodicals, and many other kinds of printing products; indeed book paper and newsprint—which has

Table 17 Components of Cost for Branches of Printing, 1954

Ratio of:	Book printing	Commercial printing	Lithographing
Cost of materials to value of shipments	34.6%	38.4%	39.9%
Production-worker payroll to value added by manufacture	54.2	47.9	45.1
Total payroll to value added by manufacture	69.4	65.4	65.1

Note: About 15 per cent of "cost of materials" represents the cost of work which the printing firm contracts out to other firms.

Source: *1954 Census of Manufactures,* Vol. II, Part 1 (Washington, 1957), pp. 27A–8, 27B–10.

a somewhat different zoning system—together make up the bulk of all paper used in printing.

The result of the system is that printers and buyers of printing in Zone No. 1 have not had to take paper prices into account as a locational factor—and Zone No. 1 includes both New York and all its principal competitive areas which are more favorably located with respect either to the cost of labor or the cost of transporting the final product.[1]

Because of this uniformity of paper prices, then, we can eliminate the cost of materials in our locational considerations. This omission leaves one other principal item in the cost structure (excluding transport costs). That is the cost of labor. In order to gauge its importance, let us consider it as a fraction of total cost minus the cost of materials, a concept which the Bureau of the Census terms "value added by manufacture."[2] A glance at Table 17 shows us that the production-worker payroll accounts for almost 50 per cent of costs (excluding materials) for the largest branch, commercial printing. The other two branches show figures that are not far different. If we then add in the wages of clerical help, supervisors, and so on, labor cost obviously becomes the dominant element in printing costs.

Labor cost is the dominant item locationally as well, because of its geographical variability. In comparison with the size of labor cost and its variability, none of the remaining items in the cost structure is of any substantial importance. Some of these are rent, taxes, depreciation, and power. They are all small fractions of cost, and even in the cases where they are geographically variable, the fractions are so small as to be swamped by the size and variability of labor cost. The principal reason for the migration of printing from the New York Metropolitan Region is the Region's top labor-cost position.

LABOR COST

Printers were among the first of the organized crafts, in New York City and elsewhere. The highly skilled nature of their work meant that they were sufficiently differentiated from ordinary labor to be able to organize, and the fact that they were literate as a matter of profession meant that they had contact with ideas of a radical sort which would lead them to organize before other groups. The earliest known American printing strike—a successful one—came during the Revolutionary War in New York City, and led to the adoption of a wage scale, but the union soon dissolved after its objective was achieved.[3] In 1809 and 1815 elaborate wage scales were won from the employers. As an indication of how old the problems of interregional competition are for New York printing, we can cite the response of the employing printers of New York to the 1815 wage scale. Their spokesman told the members of the union, the New York Typographical Society, that unless journeymen in other places would raise their prices to an equilibrium with those in New York, the differential would induce the booksellers to send their work out of the city, since the saving in labor cost would more than pay for the transportation.[4] The Typographical Society responded by adopting a resolution appointing a committee to confer with other typographical societies in the United States to persuade them to raise their wage scales to New York levels. The attempt failed.

This incident set the stage for a recurring motif in the labor history of the printing trades: the effort to eliminate interregional wage competition (and competition within metropolitan areas) by negotiation between employers and unions. For example, in 1898 the United Typothetae of America and the printing unions (compositors, pressmen, and bookbinders) entered into the so-called Syracuse agreement, in which the unions were pledged to equalize wage scales in competitive areas.[5] But the attempt failed, and it is perfectly clear why practically all such attempts have failed. Employers in the low-cost areas were being asked, in essence, to supply something for nothing. An equalization of wage scales would merely result in stopping the flow of work which they managed to divert from higher-cost areas. Labor unions in the low-cost areas were in much the same position: they saw their advantage in maintaining rates below those of the bigger metropolitan centers in order to continue the flow of work. Local organizational autonomy among both employers and unions (resulting from essentially the same competitive motivation) meant that the parent organizations could do nothing to enforce the pious hopes of those at the top of the wage pyramid.

A more recent effort was made by the National Recovery Administration (NRA) in the 1930's through its Graphic Arts Industries Code. The idea was to provide schedules of minimum hourly rates for various sizes of cities and printing firms. The schedules were designed to place important printers on a parity, at least in minimum wages, without penalizing the local small-job printer. This code vanished when the NRA was declared unconstitutional, but it is doubtful whether it would have been successful in any case, since it was an attempt to codify noncompetitive practices in an industry in which there was little interest in not competing.[6] The inclusion of these schedules in the code, however, was another indication of the concern felt by printers in principal cities over the question of differentials in wages.

Geographical wage differentials

Although the available figures on wage rates are not all that one might desire, we can make an adequate rough sketch of the labor-cost situation with their aid. The best rate to use for interregional comparisons is that for hand compositors (typesetters). Cities often undercut their competition on a single rate applicable to one sort of work, though the general level of their rates is more or less the same as that of their competitors. But we will use the hand compositor's rate as generally representative of the wage-differential picture, for several reasons. First, hand compositors form a high percentage of the printing workforce. Second, the rate is simple and unambiguous. Cylinder pressmen, for instance, outnumber hand compositors, but their rates are quite complex, depending on the kind and size of press and the number of colors printed; for cylinder pressmen New York City has 18 different rates which are not directly comparable with those of any other city.[7] Selecting an appropriate one for comparison is thus a substantial problem even for an insider. Finally, the hand compositor's rate is usually regarded as a "key" rate, setting the pattern for the "job cluster" of other rates within printing establishments.[8] It is an easy rate to bargain on, and in general, other rates have moved with it fairly closely.[9] In considering particular types of work, the hand compositor's rate could be somewhat misleading for short-period comparisons, but here we are concerned with a long-period comparison in which we need only a rough idea of what has happened to differentials.

The wage-rate figures in Table 18, comparing the years 1919 and 1955, show the following clear picture. In 1919, when New York's loss of printing was beginning to become apparent, New York and Chicago were side by side at the top of the list, and almost anywhere else in the country union minimum rates were substantially lower. If we look at the group of New York's nine principal competitors for nationally distributed printing, we can see that seven of them had minimum rates much lower than those of New York—in fact ranging from 17 to 32 per cent lower. In view of the substantial importance of labor cost, differentials of this magnitude offered considerable inducement for printing work to migrate from New York. As we have already indicated, some of these cities were close enough to New York to compete for work which would be distributed to New York's local market, as well as to national markets.

In 1955, however, the picture is much different. Without exception the other cities have raised their wages relative to New York's. Chicago and Detroit have risen above New York, and two of the three Ohio cities are pressing closely on it. The laggards—Boston and Philadelphia in particular, since they are relatively near to New York—are difficult to explain. Special local conditions may lie behind the less rapid rise in these cities: for instance, the dominance of large establishments in the Philadelphia area.

The second list in the table shows cities other than New York's

Table 18 Index of Union Hourly Minimum Rates for Hand Compositors in Large Cities, 1919 and 1955
(New York City = 100)

	1919	1955
Principal competitors (in order of 1955 rate)		
Detroit	97.2%	106.7%
Chicago	100.0	102.0
Dayton	83.3	95.8
Cleveland	83.3	93.4
Cincinnati	68.1	89.5
Philadelphia	80.5	86.7
Indianapolis	72.3	86.2
Boston	74.9	83.6
Baltimore	73.6	83.4
Other cities (in order of 1955 rate)		
Seattle	100.0	102.4
Milwaukee	72.3	97.8
Los Angeles	77.7	97.8
San Francisco	83.3	97.5
Minneapolis	73.3	94.9
Dallas	94.4	94.7
Portland, Maine	61.1	94.4
St. Louis	70.2	94.0
Pittsburgh	80.5	93.0
Kansas City	72.3	91.2
Buffalo	79.2	90.3
Atlanta	58.4	89.6
Providence	66.6	81.5

Note: These rates cover all kinds of printing except newspaper printing.

Sources: U.S. Bureau of Labor Statistics, Bulletin No. 274, *Union Scales of Wages and Hours of Labor, May 15, 1919* (September 1920); U.S. Bureau of Labor Statistics, Bulletin No. 1194, *Union Wages and Hours, Printing Industry, July 1, 1955* (March 1956).

direct competitors. In 1919 these cities had, in general, union minimum rates lower than those of New York's directly competing cities, but now the noncompeting cities are generally higher than the others. In most cases, it seems to be a question of the relative isolation of their markets: they are not printing primarily the kinds of nationally distributed products which could conceivably be printed anywhere in the country; instead, their products are either for local consumption or, if nationally distributed, are of a kind which must be printed relatively close to the originating office. Though these cities have not typically been direct competitors with New York for nationally distributed products, the fact that their wage levels are now so high compared to New York's further reduces the possibility that they will ever become so.

This table does not tell the whole labor-cost story, however, not even the whole wage story. The rates we have cited are union minimum rates, the only geographical series available over any substantial period of time. A certain fuzziness is introduced by the fact that they are minima and not prevailing rates, but the figures are adequate for the rough and ready comparisons for which we use them here. The important additional point to make, however, is that the applicability of union minima has changed considerably during the period. In 1919, among American cities, only New York was predominantly organized; almost 70 per cent of its labor force in book and job printing worked in union shops. Chicago with 50 per cent was a close second. No other city had above 30 per cent, and many were substantially less unionized.[10] There are almost no data comparing union and nonunion rates, although it is known that in general open-shop rates were, and are, lower.* By mid-century, however, the situation was quite different. In 1946 (the last year for which these figures are available) between 60 and 80 per cent of the book and job printing industry the country over was under union agreement.[11] Most of the unionized plants were con-

* As will appear presently, some open shops pay the equivalent of union rates.

centrated in the cities; so it is clear that cities in general were much more heavily unionized by 1955 than they had been in 1919. The result is that the 1919 figures *understate* the magnitude of the differentials between New York or Chicago and the rest of the cities, while the 1955 figures are much closer to the prevailing situation.

Union working rules

The differentials in wages—especially at the time of the migration of periodical printing—were dramatic and quantifiable measures of higher production cost in New York. But also of considerable importance, though less subject to specific quantification, was the question of union rules. The printing unions have classically regarded it as their prerogative to help determine the proper complement of men on the presses and the appropriate maximum rates of production, rather than leaving these to the determination of management or the result of a piece-rate system. Manning rules state how many pressmen and how many assistants are required for the operation of different types of presses. Maximum-output rules perform a similar role for typesetting. Work rules include a variety of requirements, such as a requirement that men not be shifted from one sort of a press to another—regardless of the workload in the plant or what the employer may think appropriate. This system of rules acquires interregional significance from the fact that local-union autonomy permits the establishment of different rules in different localities. For instance, Sumner Slichter noted, in 1941:

> The printing pressmen's union of New York City requires that two members of the crew on the large web presses be full-fledged pressmen, but the employers insist that only one man of the crew need be a pressman. This has been an acute issue in New York. In Chicago and elsewhere web presses are operated with only one pressman in the crew. When the presses used in printing the *Cosmopolitan, Good Housekeeping,* and *Hearst's International* were shipped from New York to Chicago, there was a reduction in the size of the crew.[12]

This difference in manning and work rules is possible even if both competitive regions are completely organized. One union group can play off against another, and the international union has nothing to say about the practice.

It was not only that rules were different—they were also differently applied. Local autonomy, especially in nonmetropolitan places, often took the form of considerably less stringent enforcement of working rules, manning rules, apprenticeship standards, and so forth. Workers regarded themselves as allied with employers in an effort to wrest work away from the metropolitan centers. A student of printing noted as late as 1942:

> [There] is a distinct tendency for less rigid enforcement of rules in smaller jurisdictions, especially in large isolated plants where the common interest of employer and labor in preserving a favorable competitive position is easily seen by both. . . . The bigger centers are handicapped both by their competitive disadvantages in rules and by the tendency towards more rigid enforcement.[13]

So far we have considered working rules entirely within a framework of unionism. But there was an even stronger locational force: open-shop competition. There were a number of open-shop centers and, in particular, a number of large open-shop firms which were competitors of New York. The largest printing firm in the country, R. R. Donnelley in Chicago, is still open-shop, with all the freedom from union rules which this implies. So are several other midwestern firms which are among the largest printing establishments in the country. Some of these large open shops have to pay the equivalent of union wage rates in order to stay nonunion, but they are not hampered by the union work rules.

TRANSPORTING THE PRODUCTS

As we have seen, labor cost—the principal geographically-variable cost component—was highest in New York and Chicago at the end of World War I. In the rest of the country, wage levels were considerably lower. Printing work of the sort which could move left the New York area in large amounts.

The place to which printing activity migrated was determined by the costs of distribution, as we will show in this section. Branches of printing in which the final product is subject to zoned rates, which vary with distance, have clustered in the Midwest. This factor is partly responsible for Chicago's continued growth as a printing center, in spite of its high wage levels. Branches of the industry subject to flat rates—rates not varying with distance—have had more freedom and have remained on the eastern seaboard to a much greater degree.

For printed matter there are a number of different transport possibilities, but the bulk of it travels at least part of the distance by U.S. Mail. In the performance of appointed rounds there is no substitute for the Post Office Department. Other organizations are not often called on to do the whole job unless bulk shipments to single destinations are possible.

POSTAGE: PERIODICALS AND CATALOGUES

When the rates for second-class material, which is chiefly periodicals, were first established in 1863, they were set at a flat rate for the entire country, like most other rates in the postal system at the time.[14] They were conceived as preferential rates for educational matter, which Congress thought should be available at low cost all over the country. But with the growth of literacy and modern methods of advertising, the character of publishers' second-class mail changed considerably. Advertising had been only an incidental feature of most magazines at the time of the rate-setting, but later many magazines came to devote as much space to advertising as they did to reading matter. Congress therefore reconsidered, and in 1918 established a two-part rate on second-class matter mailed by publishers. A flat rate was charged for the reading-matter portion of the periodicals and a zoned rate for the advertising portion. The country was divided into eight zones, and postage on the advertising portion was stepped up for each additional zone the material traveled. The importance of the two-part rate to publishers of national magazines can be judged by the fact that in 1954 the weight of the advertising portion of publications subject to zoned rates was 51.6 per cent of the total.[15]

Before the zoned system on second-class matter was created, periodical publishers did not need to consider the cost of transporting the finished product as a factor in the location of their printing activities. Newsstand distribution had only begun to get on its feet by the beginning of World War I,[16] and in that decade the publishers' principal method of distribution was by mail subscription rather than in bulk through the newsstands. (This situation still holds, but to a lesser extent.) [17] Publishers were able to keep their printing plants close to their publication offices.

The change to zoned rates meant that publishers could now no longer escape the distance factor in their distribution, no matter what channels they chose, since rates for bulk shipment by rail or motor carrier were of course distance-sensitive as well. The effect of the change to zoned rates—taken together with New York's higher labor cost—was to make locations in the Midwest increasingly attractive for the printing of periodicals. The states of Ohio, Indiana, and Illinois constitute the area of lowest-cost transport to the nation as a whole. Transport-cost calculations, taking into account the location of the market as measured by the volume of retail sales, indicate that the point of lowest cost for serving the whole United States market by land transportation falls roughly at Fort Wayne, Indiana.[18]

Time is also a factor in location in the Midwest. Large weekly periodicals with perishable contents can cut down on the time necessary to reach the whole market by locating close to its center. Many publishers now ship by rail to break-up points, at which the copies of the magazine are consigned to the Post Office. This procedure still involves distance-sensitive rates, but permits savings in cost or time over shipping by mail all the way.

Table 19 gives us some idea of the geographical distribution of the principal kinds of printing.[19] The two product branches of com-

Table 19 Geographical Distribution of Printing Receipts for Selected Products, 1954

Census region	Books and pamphlets	Periodicals	Catalogues, directories	Commercial printing and lithographing [a]
United States total	100.0%	100.0%	100.0%	100.0%
New England	11.1	7.3	6.0	6.0
Middle Atlantic total	37.2	32.1	27.4	30.7
N.Y. State and N.J.	29.8	15.6	22.3	24.6
Pennsylvania	7.4	16.5	5.1	6.1
East North Central	29.2	41.6	49.1	33.4
West North Central	5.3	3.5	5.5	7.5
South Atlantic	6.3	5.0	2.8	6.5
East South Central	5.2	b	2.9	2.6
West South Central	1.4	b	2.1	3.2
Mountain	0.8	0.7	0.4	1.3
Pacific	3.4	3.6	3.6	8.7

Note: Because of rounding, columns may not add to 100 per cent.

[a] This category of products includes periodicals, catalogues, and directories, as well as job printing.

[b] Less than one million dollars.

Source: Computed from basic data presented in Appendix D, Table D-2.

mercial printing which are most heavily concentrated in the Midwest are (1) periodicals and (2) catalogues and directories. Both of them are subject to zoned rates.

The periodicals total for the Midwest includes nearly all of the large-circulation national periodicals. The location of much of the periodical printing which is done elsewhere can be explained by the fact that the periodicals are issued by "religious, educational, scientific, philanthropic, agricultural, labor, or fraternal organizations," in the ritual formula of the Post Office Department. These nonprofit organizations are not subject to the zoned rate on the advertising portion of their periodicals. Therefore they are not concerned with transport cost in the location of their printing.

Catalogues and directories are more highly concentrated in the Midwest than any other product branch of printing. Practically half are produced in the East North Central region, and the bulk of that in Illinois. The four large mail-order houses, Sears, Montgomery-Ward, Spiegel, and Aldens, account for almost 70 per cent of the catalogue business handled by the Post Office. All four of them do their catalogue printing in the neighborhood of Chicago, shipping the catalogues in bulk via motor carrier or rail to break-up points, where they are put in the mail. (Sears had 168 of these distributing points in 1950.)[20]

The difference between the New York area and the Midwest appears more clearly if we look at some other figures. Periodicals and catalogues and directories, taken together, formed in 1954 roughly 42 per cent of the receipts of commercial printing in Illinois, the big center in the Midwest, while they formed only 14 per cent of the receipts of commercial printing in New York State.[21] That state's percentage is a rough indicator of things in the New York Metropolitan Region. We can see, therefore, that the New York Region depends very little any more on those branches in which transport cost is particularly important. It depends to a much greater extent on advertising and financial printing, especially for firms with central offices in New York City.

✓ Postage: books

The printing and binding of books and pamphlets, on the other hand, shows a lower concentration in the Midwest and a higher concentration in the Middle Atlantic and New England States than any other form of printing, as we saw in Table 19.[22] This situation is the result of the fact that most books travel by the nonzoned fourth-class rate for books, making transport cost irrelevant in the locational decision.

Table 20 shows how total book production is divided among the various branches of the book industry, branches which have different transportation needs. Starting at the top, there are general books, or trade books, sold principally through bookstores. The American Booksellers' Association estimates that its bookstore members receive about 75 per cent of their books through parcel post, at the flat rate.[23] (Packages of 70 pounds or less can travel by this rate.) Book clubs, the next branch, send all of their output by mail to individual members. Bibles and religious books, technical and professional books likewise go by mail, some to regular retail channels, some to nonbookstore distributors, and some direct to the customer. College textbooks travel to college bookstores, somewhat more by motor carrier than the branches mentioned above, but still

Table 20 Output of the Book Industry, 1952

	Percentage of total dollar output
General books, fiction and nonfiction	29
Book clubs	9
Bibles and religious books	9
Technical and professional books	11
College textbooks	7
Reference books	16
Elementary-school textbooks	13
High-school textbooks	6
	100

Source: American Book Publishers' Council, in *Postal Rates and Postal Policy of the Post Office Department,* Senate Report No. 1086, Committee on Post Office and Civil Service, 83rd Congress, 2nd Session (Washington, 1954), p. 215.

principally by mail. Reference books go either to bookstore channels by mail, or by direct mail to the customer.

These branches we have so far named account for 81 per cent of the total output. As we have seen, the books in each branch travel principally by mail at the nonzoned book rate. Consequently we would expect to find less pressure for a midwestern location in most book printing—an interpretation which fits the figures presented in Table 19.

The remainder of the book industry's output is high-school and elementary-school textbooks. These ordinarily travel in large lots by motor carrier to state textbook depositories and school systems.

Motor carrier rates are of course distance-sensitive. Although adequate statistics are not available, industry spokesmen indicate that this form of book printing tends to concentrate much more in the Midwest than other book printing, but as the table shows, those two branches account for only 19 per cent of total book output.

Postage: general printed matter

The residual after we have considered these types of printing beggars easy description; so we just give it the title of general printed matter. Much of it is printing for a local or regional market: business forms, menus, programs, and so on. But that part of it which is produced for the national market can be located in the New York Metropolitan Region as far as transport cost is concerned. The most important single category of this residual national-market printing is probably direct-mail advertising, although figures are completely lacking on this point. For direct-mail advertising and similar types of printing, distribution is principally by mail to individual addressees. Since 1928 there has been a special bulk third-class rate for this material. Identical pieces of third-class matter mailed at one time in quantities of at least 20 pounds or 200 pieces, pre-sorted by state and city, travel at a flat rate the country over.[24]

This special circumstance allows direct-mail advertising and similar individually addressed matter to be printed in the New York Metropolitan Region even though it is destined for national distribution. There is no transport-cost advantage in having it printed somewhere else, and considerable reason for it to be done in close proximity to the place where the material is created.

The importance of this flat rate for printers is indicated by an estimate by the Printing Industry of America, the national printing trade association:

> Material prepared for third-class mail represents the bread and butter of business of most commercial printers. We have estimated that as much as one-third of the dollar volume of the printing industry is designed for distribution by third-class mail.[25]

The potential importance of the transport-cost factor is indicated by the further statement that

> If we examine the cost of mailing we find that the postage charge is often as great or greater than the cost of manufacturing the mailing itself.[26]

But as long as rates remain unzoned (and there seems to be no move to change this feature of the rate structure), transport cost, although potentially of great importance, can have no influence on the location of advertising printing.

SUMMARY

The cost situation in printing viewed as a whole is this: New York is still the city with the highest printing costs in the East. A considerable volume of printing is still done in the New York Metropolitan Region, however, primarily because it is either printing for local consumption or printing in which consultation and deadlines are dominant—or both.

Long-run printing which is primarily for the national market has largely left the New York Region to seek lower-cost locations elsewhere. The particular area to which it has migrated has been determined by distribution costs. In periodical printing and catalogue printing, the work has been forced to the center of the market, to minimize the time and money costs of serving the national market with a distance-sensitive rate structure. Communication problems between printer and publisher have been solved by having the consultation-oriented stages of typesetting and engraving in New York and the routine stages in the Midwest. Book printing, on the other hand, has not been forced to the center of the market, but has migrated to low-cost areas on the eastern seaboard. General printed matter for the national market—especially where it involves deadlines and much consultation between issuing office and printer—has been able to stay in the New York Region to a greater extent than the other work because of the nonzoned rate applicable to direct-mail advertising and similar materials.

5

The Next Twenty-five Years

In order to assess the future of printing and publishing in the New York Metropolitan Region, we must first make an estimate of what the national employment in printing and publishing will be, and then apply the results of our earlier analysis to see what is likely to happen to the Region's share of that employment. We will be concerned with the years 1965, 1975, and 1985, in common with the other studies in this volume.

NATIONAL EMPLOYMENT

For purposes of projection, printing and publishing can be divided into three sectors: (1) newspapers, including both the publishing and printing aspects; (2) publishing, which means primarily the preparation of the content of periodicals, books, and other published products besides newspapers; (3) printing, which covers the mechanical production of the publications prepared by the publishing sector and of all sorts of miscellaneous printed materials.

THE NEWSPAPER SECTOR

Daily newspapers, which account for over 90 per cent of total newspaper circulation, have been closing down or merging at such a rate as to cause doubts about the future growth of the industry. Between 1929 and 1944 there was a net decrease of 200 in the dailies of the United States—the number dropping from 1,944 to 1,744. But in more recent years there has been a net increase, because the continuing suspensions have been more than offset by new papers.

These new papers have sprung up mostly in fast-growing communities of less than 10,000 population. From 1944 to 1954, the number of cities with dailies increased by 52—from 1,396 to 1,448—and the net rise in the total number of dailies was 19.[1] The most recent figures show a net increase of 20 in only one year, from 1957 to 1958.[2]

Can the upward trend be expected to go on? We think so. Because of the phenomenon of suburbanization, which we examined in an earlier chapter, it seems reasonable to expect more new papers to arise to serve communities becoming large enough to support them, or to serve regions which have become market areas for new shopping centers. (Since 1952 the New York Metropolitan Region has added one daily and lost none.) Besides, the changes in technology which can be expected over the next 25 years can be counted on to make it easier to start small papers by reducing the capital cost and hence the risk associated with it.[3]

On the other hand, relatively few newspapers are likely to discontinue operations in the future. In the words of one of the leading students of journalism:

> Outright suspensions will not be so numerous during the next 20 years as during the last three decades, for the simple reason that there are not many unprofitable papers left. The relatively few "marginal" dailies in this country today are principally at either end of the spectrum—a few in places too small to support a daily . . . and a few more in large metropolitan areas where there is excessive competition for essentially the same kind of readers. Most other papers today are making a reasonable profit and are basically sound—first, because of the essential nature of the commodity they sell, and, second, because of the non-competitive position they hold.[4]

What significance does this discussion have for employment trends? Unless new dailies emerge much faster in the next 25 years than anyone now expects, the difference which they will make in the industry's employment will be relatively slight. The important thing is that a contrary trend is *not* taking place: the country is not being served from an increasingly small number of centers by fewer and fewer papers. If it were, we could feel fairly sure that the substantial economies that come with large-scale operations would lead to a fall in newspaper employment for the same predicted level of circulation. The possible development of "national newspapers" will be discussed later in this chapter; suffice it to say at the moment that we do not expect such papers—if they arise—to have a diminishing effect on newspaper employment.

With these remarks on the number of newspapers as background, let us then turn to some more precise indications of the demand for newspapers. Trends in newspaper circulation are, in general, closely related to trends in population. But there are also changes in *per capita* consumption of newspapers, and if we are projecting some distance into the future, these changes are likely to be significant.

The middle column of Table 21 shows that per capita sales of newspapers moved upward during the war period but declined a little from 1947 to 1954. Can we expect this decline to continue? Such a change cumulated over 25 years could make a substantial difference in our estimates of demand.

Table 21 Actual and Projected Newspaper Sales, United States, Selected Years

	Total copies per week [a] (in thousands)	Per capita	Per person 15 and over
1929	304,013	2.497	3.553
1939	313,735	2.397	3.209
1947	386,165	2.692	3.657
1954	418,874	2.599	3.669
1965	513,600	2.641	3.688
1975	596,400	2.673	3.705
1985	671,100	2.685	3.723

[a] Figures were computed by multiplying average circulation per issue by frequency of issue per week.

Source: 1929–1954 figures from Bureau of the Census, *Statistical Abstract of the United States: 1958* (Washington, 1958), p. 523.

A clue to this postwar decline in per capita newspaper sales lies in the size of the newspaper-buying public relative to the total population. The years since the war have seen a substantial increase in the relative importance of the youngest age groups. In 1945, for instance, 25.2 per cent of the nation's population was under the age of 15; by 1956 this proportion had increased to 30 per cent.[5]

But the relative growth of the youngest age groups now appears to be over. The population estimates which we use for projection purposes, for instance, indicate that the percentage of the population under 15 will fall to 28.2 per cent of the total in 1965, and to 27.9 per cent for both 1975 and 1985.[6]

The projections in Table 21, of course, would not materialize if advertising in newspapers were to fall off. For newspaper demand comes in two pieces: the demand for the papers by the consumers, and the demand for the advertising space which reaches the consumers. Revenue from advertising is more than two-thirds of total newspaper revenue, and if advertisers shift to other media, the consequences for newspapers can be quite as serious as a radical shift in circulation patterns.

An analysis of the demand for competing advertising media can be an extremely complex affair. Here we are relying primarily on the postwar behavior of newspaper advertising on which to base our expectations. The physical amount of advertising placed in newspapers—advertising linage—increased by 45 per cent from 1947 to 1956.[7] This postwar picture is in considerable contrast to the doldrums in which newspaper advertising had been since 1929. One factor pointing to the continued growth of newspaper advertising is the introduction of color. Color advertising places the newspaper in a better competitive situation with respect to other advertising media, and advertisers are responding: color ad linage more than doubled from 1952 to 1957.[8]

The remaining problem in the newspaper sector is to transform projections of newspaper sales into projections of newspaper employment. First we must translate future newspaper sales into total output—taking into account not only the number of copies but also the size of each copy. The best measure of output is the amount of paper consumed in the printing of newspapers. The tonnage figures are available for 1947 and 1954, and these are shown in Table 22.

Table 22 Actual and Projected Newsprint Consumption, United States, Selected Years (in short tons)

	Total	Per thousand copies per week
1947	4,162,848	10.78
1954	5,262,177	12.56
1965	7,278,000	14.17
1975	9,423,000	15.80
1985	11,832,000	17.63

Source: *1954 Census of Manufactures,* Vol. II, Part 1 (Washington, 1957), p. 27A–20; *1947 Census of Manufactures,* Vol. II (Washington, 1949), p. 353.

By combining them with the sales figures given in the preceding table, we arrive at newsprint consumption per unit of sales for 1947 and 1954, shown in the second column of Table 22. The consumption per unit increased by almost 16 per cent between the two dates; papers have been getting larger on the average. It is questionable whether we can expect this increase to continue at quite such a high rate. As a rough adjustment, then, let us assume that the future rate will be half of what it was from 1947 to 1954. On this basis we can calculate total newsprint consumption for our target years. The figures are presented in the lower part of Table 22.

The next stage is to relate newspaper output, as measured by paper consumption, to the number of workers employed in the industry. As newspaper output has grown, there has been a change in the distribution of newspaper employment between production workers (primarily those involved in printing) and nonproduction workers (primarily those involved in publishing). Nonproduction workers decreased by almost 11 per cent in relation to output from 1947 to 1954. The decrease is the result of the fact that publishing inputs can be regarded as a more-or-less fixed expense for newspapers. Circulation can be expanded without adding to the publishing inputs. Production workers, on the other hand, have expanded in proportion with output. This growth is a compound of two effects: the often-cited slow rate of productivity increase in newspaper printing, and the relatively rapid rate of employment growth in small newspapers where labor requirements per unit of output are higher than they are in larger papers.

For projection it seems safe to assume that the number of newspaper publishing workers will continue to grow more slowly than output, as we have done in Table 23. For production-worker employment, however, we can expect a performance rather different from that of the past. Although productivity change has been slow for decades in the newspaper industry, there is on the horizon a series of innovations which should cut down considerably on labor requirements in every stage of newspaper production. In typesetting there are new machines which set type photographically, using no metal at all. They show promise of being especially useful in advertising composition. Some papers are already using the teletypesetter to set press association stories in type by remote control. Faster photoengraving processes are being introduced. There are

Table 23 Actual and Projected Newspaper Employment, United States, Selected Years

	Total	Nonproduction workers	Production workers
1947	234,375	116,259	118,116
1954	297,799	132,858	148,941
1965	363,000	157,000	206,000
1975	417,000	176,000	241,000
1985	452,000	191,000	261,000

Source: *1954 Census of Manufactures,* Vol. II, Part 1 (Washington, 1957), p. 27A-4.

potential developments of new sorts of printing plates which will allow newspapers to short-circuit the stereotyping process. Various developments on the presses are in the offing, and, finally, the "mail room," where papers are bundled, sorted, and loaded, seems to be heading for more mechanization.[9]

It would be a mistake to assume that the newspaper business is going to be transformed overnight. In most cities, newspapers are monopolies and do not have competitive pressure to introduce new methods. Then there is labor's traditional resistance to technological change in the newspaper industry. Finally, there is the fact that many of the most dramatic of these innovations are still in the talking stage, and yet others are only working models. Consequently we do not think that an immediate change in productivity patterns is likely—and for this reason we predict no change in productivity by 1965, our first target year. By 1975, however, we expect the change to be well underway, and by 1985 even further along. On this basis we assume a 1 per cent annual increase in the productivity of production workers for the decade 1965 to 1975, and an annual increase of 1.5 per cent from 1975 to 1985. These assumed rates of productivity change, when applied to the projections of output given in Table 22, yield the figures on employment of production workers which are entered in Table 23.

The production workers and nonproduction workers, added together, give us the total newspaper employment shown for the three target years.

The publishing sector

The publishing sector, as we define it here for prediction purposes, includes the publishing workers in the industries issuing periodicals, books, and miscellaneous published products.

Periodicals. Like newspapers, periodicals have shown increases in per capita consumption since 1929, as Table 24 shows, but the increase from 1947 to 1954 was not as rapid as those of earlier periods.

The chief competitor for magazine circulation and advertising

Table 24 Actual and Projected Annual Sales of Periodicals and Books, United States, Selected Years

	Periodicals			Hardbound Books			Paperbound Books		
	Total (thousands)	Per person	Per person 15 and over	Total (thousands)	Per person	Per person 15 and over	Total (thousands)	Per person	Per person 15 and over
1929	4,206,934	34.55	49.17	n.a.	n.a.	n.a.	n.a.	n.a.	n.a.
1939	5,899,080	45.07	60.34	n.a.	n.a.	n.a.	n.a.	n.a.	n.a.
1947	6,965,253	48.56	65.97	277,450	1.93	2.63	209,766	1.46	1.99
1954	7,952,865	49.34	69.67	302,255	1.87	2.65	439,062	2.72	3.85
1965	10,572,000	54.47	75.91	373,000	1.92	2.68	912,800	4.70	6.55
1975	13,209,000	59.19	82.06	436,000	1.95	2.71	1,378,000	6.17	8.56
1985	15,993,000	63.99	88.72	494,000	1.97	2.74	1,745,000	6.98	9.68

n.a. = not available.

Sources: Computed from the following: for past book sales, *1954 Census of Manufactures*, Vol. II, Part 1 (Washington, 1957), p. 27A–12; *1947 Census of Manufactures*, Vol. II (Washington, 1949), p. 358; for past periodical sales, *1954 Census of Manufactures*, Vol. II, Part 1, p. 27A–16 (figures were computed by multiplying average circulation per issue by frequency of issue per year); for population statistics, see our Appendix D, Table D–3.

revenue has been television. Studies of readership among owners of television sets show that magazines have suffered directly because of television.[10] The effect can be traced to some extent in circulation figures as well; during the freeze on the licensing of television stations from 1946 to 1952, magazine circulation increased by 8 per cent in cities of more than 100,000 population but increased by 28 per cent in places of less than 100,000 population, which had few television stations. Since that period, with television mushrooming in the smaller places, the situation has been reversed; circulation in the large cities has grown faster than circulation in smaller places. The groups of magazines which are most closely competitive with television are the ones which have suffered most: "romance" magazines, sensational picture magazines, and the like. Those leading the magazine parade in the rate of circulation growth have been the high-brow magazines, closely trailed by magazines appealing to special-interest groups: hobby and do-it-yourself magazines and trade publications, for instance. The over-all picture in magazine circulation is one of increasing demand, however, in spite of losses or slow growth among particular groups. For magazines as a whole, circulation per capita among the adult group has shown a substantial rise, although per capita consumption for the whole population has moved upwards only slightly because of the relative increase in the population under 15 which we have already discussed.

The consequences of television for advertising revenues have been rather more serious. Periodicals are in roughly the same position as newspapers in that the advertiser pays most of the cost. Forty-eight general-circulation magazines have gone out of business between 1948 and 1958—some with circulations in the millions, like *Collier's* and *Woman's Home Companion*.[11] *Collier's*, in fact, had the highest circulation in its history at the time it folded. What has apparently happened is that the competition from television has picked off the weakest magazines in each group (general consumer magazines, women's magazines, and so on), as advertisers have withdrawn their support in favor of TV. Figures on magazine advertising linage show a sharp drop in 1949 when TV advertising was introduced, and until recently the linage stayed below the 1947 level.[12]

The result has been that magazine publishers in general have tended to move away from direct competition with TV, in an attempt to exploit areas of consumer interest and advertising appeal where the ravages of TV will not be felt so strongly. In doing this they are taking their cue from the circulation increases in magazines which are by nature less competitive with TV. For example, the content of magazines has shifted since the war; in particular, their content of fiction—the item most obviously competitive with TV—has dropped dramatically. Even magazines appealing to special groups (housewives, do-it-yourself addicts) have changed their contents in order to concentrate more heavily on their specialties.[13]

What inferences can we draw from this discussion for our projections? The chief point is that magazines are not sitting ducks for the competition from TV. Apart from a handful of large, general-interest magazines they cater to specialized groups of readers and to advertisers who want to appeal more effectively (or more selectively) to these specialized groups than they could through the medium of television. Magazines seem to be moving in the direction of exploiting this advantage further. Per capita readership has continued to rise, and the advertising picture in the most recent years has shown a considerable recovery. Our inclination, then, is to project a continuation of the rise in per capita readership of periodicals.

Books. Our data on book consumption indicate only a slight rise in readership of hardbound books among the adult population from 1947 to 1954, but very close to a doubling of readership for paperbound books, as Table 24 shows. Although originally the paperbacks were principally a vehicle for reprints of popular successes from the hardbound field, in the last five or six years more and more original works have been published in paperback form, or in simultaneous hardbound and paperbound editions. Among the reprints, an increasing number are books of the sort which had previously been restricted to hardbound publication. Many scholarly works have achieved remarkable sales in paperback format. Paperbacks have tapped an entirely new market; there are something like 3,000 retail outlets for hardbound books, and roughly 100,000 for paperbacks.

In Table 24 we project future per capita consumption of hardbound books and paperbacks separately, since the trends are quite different. For hardbound books, we are projecting the current small rate of increase in adult per capita consumption. Since paperbacks are in their early stages, however, their recent rate of growth has been much higher than one can expect to continue. Projecting the current rate of increase would have the adult population consuming 72 paperbacks per year by 1985. Consequently we assume that the 1954–1965 rate of growth will be half of the past one, and that the rates for the next two decades will each be half of that of the preceding decade.

Projecting employment. Turning these future sales levels into employment figures involves some rather difficult questions. Part of the publishing work force varies markedly in relation to the number of copies sold—for example, magazine employees who are engaged in subscription fulfillment work, such as accounting and billing. These sales-related employees have probably increased in the postwar period, though increased mechanization of routine clerical operations must have prevented them from increasing as fast as the sales.

The other component of publishing workers bears no close relation to sales. This group is composed of editors, manuscript readers for book publishers, staff writers for the magazines, advertising solicitors, and so on. This category of workers is more closely related to the number of distinct products put out by the industry than to the number of copies sold.

There is no way of moving surefootedly from these impressions to hard projections of employment in the publishing end of magazine and book production. Our disposition is to assume a slow rate of employment growth notwithstanding the projected considerable increase in magazine and book sales; the realism of this assumption is fortified by the experience of the postwar years. Accordingly, with a projected increase in sales of 33 per cent in periodicals, 24 per cent in hardbound books, and 108 per cent in paperbound books from 1954 to 1965, we nonetheless project an increase in publishing personnel of 10 per cent. In subsequent decades, the publishing employment increase is also likely to be well below the sales increase. On top of this, provision is made for the continued rapid expansion in employment of that small but ebullient and indefinable group called "miscellaneous publishing." Its 50 per cent increase from 1947 to 1954 is extended in our projections from 1954 to 1965 as an additional 50 per cent increase, and is tapered off from there. The resulting employment figures for the entire publishing sector emerge as follows:

1965	103,000
1975	116,000
1985	128,000

THE PRINTING SECTOR

The printing sector, as we have carved it out here for projection purposes, consists of all the printing and servicing stages which go into the production of any printed matter—except for printing done in newspaper plants. Part of this output, such as periodicals, books, maps, and sheet music, is eventually purchased by the public. Part is produced for intermediate use by other industries, for example, business forms, calendars, advertising materials, and catalogues.

We have already found that in the postwar period there has been a slightly rising per capita consumption of both periodicals and books. How do these figures compare with paper consumption in the printing sector—our measure of that sector's output? The middle columns of Table 25 show the increase in consumption of paper used for periodicals, books, and the miscellaneous products of the publishing industry from 1947 to 1954. On a per capita basis

Table 25 Actual and Projected Paper Consumption in the Printing Sector, United States, Selected Years (in short tons)

	Used for all printing		Used for publications		Used for other products	
	Total	Per 100 persons	Total	Per 100 persons	Total	Per 100 persons
1947 . . .	3,291,182	2.294	1,759,223	1.226	1,531,959	1.068
1954 . . .	4,283,809	2.658	2,313,443	1.435	1,970,366	1.222
1965 . . .	6,502,000	3.350	....	....	....	....
1975 . . .	9,228,000	4.135	....	....	....	....
1985 . . .	12,755,000	5.103	....	....	....	....

Source: *1954 Census of Manufactures,* Vol. II, Part 1 (Washington, 1957), pp. 27A–20 and 27B–21; *1947 Census of Manufactures,* Vol. II (Washington, 1949), pp. 360 and 375.

this increase was 17 per cent. We cannot say whether this rise was due to the considerable growth in miscellaneous publishing, as it might have been, or to a growth in the average size of periodicals and books; probably both were involved. For the wild variety of other printed products besides publications, such as stationery and calendars, the increase in per capita consumption has been 14.4 per cent.

We assume that these patterns will continue into the future, and on this basis we project the 1947–1954 rate of increase in per capita paper consumption for our three target dates. The results of the computations are entered in Table 25. The projections are consistent with our projected increases in per capita sales of books and periodicals (the principal products of the publishing sector), and there seems to be no reason for assuming any decline in the rate of increase for paper used for other products.

What about productivity in the printing sector? The output per man, as measured by the paper inputs, increased by approximately 16 per cent from 1947 to 1954, or at an annual rate of about 2.2 per cent, a rate somewhat lower than that for the economy as a whole. As we have already indicated with respect to newspaper production, however, there are potential advances in productivity which we can expect to start taking effect during the decade 1965–1975, and which should have even more force in the decade following that. Consequently we will assume the continuation of the past rate of productivity change in the industry (2.2 per cent) only until 1965. From 1965 to 1975, we will assume the annual rate to be 2.5 per cent, and from 1975 to 1985, we will assume it to be 3.0 per cent.

When we apply the productivity increases to the expected output figures, we arrive at the following projection of national employment in the printing sector:

1965	523,000
1975	581,000
1985	598,000

Now we are ready to add together the employment projections for the three sectors in order to forecast total employment in printing and publishing in the United States. The actual total in 1947 was 715,120, and in 1954 it was 804,382. For our three target years we estimate the totals as follows:

1965	989,000
1975	1,114,000
1985	1,178,000

EMPLOYMENT IN THE REGION

An intricate set of factors—some adding to the competitive strength of the New York Metropolitan Region, some detracting from it—promises to leave the Region's relative position in the nation's printing and publishing industry almost unchanged over the next few decades. Once again we must examine each of the three sectors.

THE NEWSPAPER SECTOR

A useful guide in attempting to move from projections of national newspaper employment to projections of the Region's newspaper employment is the increase of prospective readers in the Region relative to the increase in the nation, a guide which can be based on comparative population increases. On this basis, one would be disposed to anticipate a decline in the Region's share of newspaper employment.[14] In addition, the increasing urbanization of the rest of the nation has to be taken into account. At present, the Region's per capita consumption of newspapers is somewhat higher than that of the nation. That margin should be reduced somewhat as the urbanization process continues. The effects of the Region's declining share of national population and the increasing urbanization elsewhere are reflected in the figures presented in Table 26, where we see that the Region's share of national newspaper employment dropped from 12.5 to 10.8 per cent between 1947 and 1954. Our expectations about the future trends are incorporated in Table 27.

Table 26 Printing and Publishing Employment, United States and New York Metropolitan Region, 1947 and 1954

	1947			1954		
	U.S.	NYMR	Ratio, NYMR to U.S.	U.S.	NYMR	Ratio, NYMR to U.S.
Printing and publishing, total	715,120	164,745	23.0%	804,382	167,227	22.8%
Newspaper sector	234,375	29,433	12.5	281,799	30,439	10.8
Publishing sector	89,941	49,457	55.0	85,089	43,168	50.8
Printing sector	390,804	85,855	22.0	437,494	93,620	21.4

Source: National data are from *1954 Census of Manufactures,* Vol. II, Part 1 (Washington, 1957). Regional data are the sum of federal data for the Standard Metropolitan Area and our estimates for the five additional counties of the Region; these estimates are based primarily on county data from state departments of labor.

Table 27 Projections of Printing and Publishing Employment, New York Metropolitan Region

	1965		1975		1985	
	NYMR employment	Ratio, NYMR to U.S.	NYMR employment	Ratio, NYMR to U.S.	NYMR employment	Ratio, NYMR to U.S.
Printing and publishing, total	200,000	20.2%	221,000	19.8%	233,000	19.7%
Newspaper sector	36,000	10.0	39,000	9.4	41,000	9.0
Publishing sector	52,000	50.0	58,000	50.0	64,000	50.0
Printing sector	112,000	21.4	124,000	21.4	128,000	21.4

What of the possibility of printing newspapers "by remote control," as described earlier for periodicals? Developments of this sort are already a reality to some extent. The *Wall Street Journal* prints four regional editions in addition to its New York edition, using teletypesetter machines to set identical type at various points in the country. On two occasions—in 1945 and again in 1954—the *New York Times* produced complete newspapers in its New York office and transmitted the pages to San Francisco by facsimile, where they were printed and distributed.

The spread of "national" newspapers published in this fashion seems to be in the cards within the period of our projection, and New York seems the logical center, for reasons which we mentioned in our chapter on periodical publishing. But this will not alter the essential nature of newspapers. They will remain a local-market industry, with each paper maintaining its employees at the heart of its news sources and its reading public. National papers, we assume, will appear as adjuncts of going enterprises, as has already happened in the case of the *Wall Street Journal*. Employment in the New York Metropolitan Region on national newspapers will probably be quite small, since they will be able to use material already prepared for their New York editions. The development will also generate employment where the regional editions are published, since much of the content of the paper—both advertising and reading matter—will have to be local. Accordingly, the spread of national newspapers, as we see it, does not require us to make any adjustment in the figures.

✓ The publishing sector

As we have seen, publishing is concentrated in New York for two primary reasons: (1) the Region is the communications and information center of the nation and (2) it possesses the principal concentration of the specialized labor force which publishing requires. Publishing, as we have seen, is dependent on an atmosphere of close and continuing personal communication, and it is difficult to see how technological advance could do anything to break up New York's tight-knit complex, even if closed-circuit TV turns out to be a replica of e. e. cummings' "lookiesoundiefeelietastiesmellie."

To be sure, between 1947 and 1954, the Region's publishing sector declined somewhat as a percentage of national publishing activity. But year-to-year employment figures in publishing show such erratic fluctuations that we are not inclined to regard the change between 1947 and 1954 as suggesting a trend. Long-run tendencies in the publishing field prior to 1947 have been to increase its concentration in the New York area, and our expectation for the future on the basis of our study is that New York's share will remain constant. The figures in Table 27, showing publishing employment in the New York Metropolitan Region for 1965, 1975, and 1985, are based on this assumption.

✓ The printing sector

Most of the printing work which can be done on better terms in areas outside the New York Metropolitan Region has already left the Region. An exception is book printing, a category in which more emigration may yet develop. Book printing, however, represents a relatively small fraction of all printing and publishing in the Region, and it would take quite a substantial loss in book printing to make much change in the total.

Going back to Table 26, we see that the Region's share of the nation's printing employment changed very little between 1947 and

1954. What could lead to substantial loss of printing from New York? A tendency toward further improvement of transport and inter-city communications would have such an effect. It is doubtful that much improvement can be expected in truck and rail times between New York and printing centers that have lower costs. If air freight comes into heavy use for printed products, New York may lose more of its printing business, that is, provided labor-cost differentials continue wide enough. But in any case, the sort of work in which the crucial factor is the transport of the finished product from the plant to New York is only a part of the total picture. As we saw earlier, a great proportion of the work which is still done in New York depends on the city's complex of subcontractors, many of whom have to consult with the customer. It is a question not simply of moving printed materials from plant to distribution point, but of interplant shipments, and of consultation, correction of proofs, and so on at each stage of the production process—typesetting, engraving, presswork, and binding. The more hands there are in the production of a tailored and speed-demanding printed product originating in New York, the more likely it is that New York or its immediate surroundings will be an optimal location for it.

There are on the horizon certain innovations which one would think—at first glance—might reduce to some extent the locational importance of communication. In particular, facsimile transmission might be introduced in the printing industry. Proofs of text material, for example, might be sent by a facsimile process from the printing plant to the issuing office, where they could be reviewed and sent back by facsimile to the printer, complete with corrections. Certain forms of relatively simple printed matter—single-color reproduction of a text with few illustrations, for example—could be affected by such a development.

The critical factor, however, is that the buyer of the printing wants to see his final product as it will appear to the reader, and there are many problems of detailed personal inspection of color and texture (along with other aspects) which facsimile probably will not be able to handle well during the next 25 years. Prediction of technological change is necessarily risky, however, and facsimile may develop fast enough to be a substantial factor within that period.

Set against more rapid transport and better communications, there is another trend that must be taken into account. As we have seen in the discussion of labor cost, New York's cost disadvantage has been steadily shrinking since 1919. Consequently there is less and less reason for printing to move out of New York, especially when it has to do so at the cost of increasingly expensive devices for improved communications.

One factor which we have been holding constant throughout this analysis is New York's status as a printing consumer. A major reason for New York's continued predominance as a printing center has been the presence of many of the largest consumers of printing: advertising agencies, public relations firms, central offices of corporations, foundations, trade associations, and so on. If such activities were to begin emigrating from the Region, then the printing which is tied to them would also go.

In other parts of the New York Metropolitan Region Study this problem is being explored in detail. The preliminary indications are that New York is not likely to lose ground as a center for these types of office activities and may even gain. Jet planes, by making possible faster contact with the rest of the country, may contribute to a further centralization of activities which depend on close personal contact; top management, for instance, will be able to locate in New York and still make frequent visits to their factories in other parts of the country.

Balancing the various forces at work, we assume that the Region's share of the nation's printing employment will be the same in 1965, 1975, and 1985 as it was in 1954. The resulting employment figures are presented in Table 27 along with those of the newspaper and publishing sectors.

INSIDE THE REGION

Our final concern is to project the distribution of printing and publishing jobs inside of the New York Metropolitan Region. In Tables 3 and 4, presented in Chapter 1 of this study, we saw that New York City's share of the Region's printing and publishing employment has fallen considerably during the century. By 1954 the city's share was down to about 75 per cent, and between 1947 and 1954 there was even a slight decrease in the city's absolute number of jobs.

The city's decline relative to the rest of the Region was the net result of different patterns of behavior in the various sectors of the industry. Newspapers redistributed their employment in response to the higher relative growth of newspapers outside of New York City. Publishing activities, on the other hand, clung strongly to Manhattan for all the reasons discussed earlier. Printing tied to customers in the central business district did not leave the city, but there was some shift of other printing to outlying counties of the Region.

But these are impressionistic observations. Reliable employment data by parts of the Region do not exist for the sectors of the industry which we have used up to now.[15] All one can do in projecting intra-regional trends is to appeal to aggregated printing and publishing figures, despite the heterogeneous nature of the activities they cover. We will use our projections of employment in the Region, and assume that employment in New York City will decline relative to the Region at the 1947–1954 rate. Under these assumptions, the city would increase its jobs from 127,000 in 1954 to 139,000 in 1965 and 142,000 in 1975, after which there would be a decline to 136,000 in 1985. On the same basis, jobs in the Region outside New York City would climb from 40,000 in 1954 to 61,000, 79,000, and 97,000, respectively, on the three succeeding dates. New York City's printing and publishing employment seems in for considerable decline relative to the rest of the Region, but it also seems likely to experience a resumption of absolute growth, which would reverse its 1947–1954 decrease, primarily because of the anticipated expansion in publishing, a highly city-oriented activity. Our inclination, after experimenting with various possibilities in the projected distribution of employment in the Region, is to suspect that the projected absolute growth for New York City is probably a sound assumption, and that the extent of the growth may well prove greater than the figures suggested by this mechanistic technique of projection.

All in all, the position of the New York area as the nation's dominant metropolis in the creation and exchange of ideas and images in printed form seems assured for the decades ahead.

ELECTRONICS

By

James M. Hund

Appendix E, Electronics Projections—Sources and Methods, begins on page 355.

Notes to Electronics begin on page 364.

Introducing an Industry*

Before man knew for sure that electrons existed, he learned how to use them as electrical currents by forcing them to flow from atom to atom in a copper wire, thereby making them do jobs like turning a wheel and lighting a street. Thus he created the electrical industry. Later, man discovered more about controlling electrons. He found that in a vacuum tube their flow could be precisely controlled and a very small voltage greatly amplified. For example, he could receive a Beethoven symphony in the form of feeble, vibrating radio waves, lead it from a rooftop antenna into the path of the stream of leaping electrons in the tube (or a series of tubes), and thus translate it into a strong current sufficient to stir a loudspeaker into booming sound. By elaborations of the process he could translate a distant airplane into a moving spot on a screen; a home run in the stadium into a home run in the living room; and a set of instructions on magnetic tape into automatic operations within a factory. By the 1940's it was said—and with a little imagination could almost be believed—that the electron tube could "talk, hear, see, feel, taste, smell, sort, count, regulate, measure, calculate, and even remember."[1]

Possessing such an array of talents, the electron tube, along with later devices that do the same work, has become the basis of a new industry, which, although decidedly electrical, uses streams of electrons not for routine tasks like lighting, heating, or substituting for human muscle, but for taking over some of the functions of the

* The author wishes to thank Benjamin Chinitz for valuable help in the preparation of this study.

human brain.[2] This new industry has acquired a bigger dollar mark than its parent, the electrical industry, and now is usually placed in a separate category and called by a different name—electronics.

The electronics industry is affecting, or is about to affect, human existence so pervasively that we are more justified in calling this the Electronic Age than the Atomic. One surmises that atomic energy is in about the same stage of development that electronics was in the 1920's, and that during the rest of the century the two industries may move closer and closer together, as the word *nucleonics* already suggests. The electronics industry has invaded the consumer's budget with radio, television, and high fidelity; the businessman's budget with data-processing equipment and automated machinery; and the national defense budget with radar and a myriad of control devices for the operation of aircraft, missiles, artillery, and naval vessels. It is now invading the scientific laboratory and the hospital. In the process, it has come to occupy an important position among the nation's manufacturing industries. Between 1939 and 1954, employment in electronics production, as nearly as we can tell, increased 400 per cent, from 81,000 to 410,000.[3] In the same period, employment in all manufacturing was increasing 65 per cent, from 9,500,000 to 15,700,000.

As new manufacturing industries are born and develop, they not only alter the nation's patterns of consumption and methods of production, but also affect the location of the nation's jobs. In their infancy they are attracted to one kind of area and repelled from another. As they grow, their locational preferences may alter. Electronics affords a classic example of a new industry which has made locational shifts during its adolescence. It is an industry on the wing.

Scope of the inquiry

In this study our concern with electronics is the concern of an economist. We do not pretend to explore thoroughly the scientific, engineering, or military sides of electronics—only enough to throw necessary light on the economic side. Our subject is the manufacturing industry; and more particularly where it has been located and why; and still more particularly how much of it has been, and will be, located in the nation's most populous metropolitan area.

The industry provided about 76,000 jobs in the New York Metropolitan Region in 1954, the latest year for which United States Census figures are available.[4] This was somewhere around 19 per cent of the nation's employment in electronics. Statistics furnished by state governments indicate that in 1956 the New York Metropolitan Region's employment had risen to about 98,000. This was only 5 per cent of the Region's employment in all manufacturing; but 5 per cent understates the weight that the industry can swing in the Region. For we can have no doubt that American electronics production will continue to grow at an extraordinary rate—much faster than manufacturing generally. And therefore the Region's trend in total manufacturing employment during the next quarter-century will be considerably influenced by the extent to which it attracts electronics production.

The electronics industry is remarkably difficult to quantify and classify. In the first place the speed with which it grows and changes from year to year renders the statistics seriously out of date before they can be scraped together and printed. Until World War II, electronics was almost synonymous with radio; today radio is only one portion. Even the electron tube is already being replaced in many products by the smaller transistor, in which patterns of communication are imposed on electrons not as they leap through a vacuum but as they traverse a tiny lump of germanium or silicon. But even if the statistics were up to the minute, the industry would still be elusive because its products are not officially classified in a neat "electronics" bundle; they fall into many different industries and are not easily separated from the nonelectronic products.

"The first thing to be grasped about electronics as a business, A.D. 1957, is that this is an industry running on three different clocks," wrote William B. Harris in the April 1957 number of *For-*

tune. "One part of the industry is already 'old' enough to have seen severe competition, strong forces toward concentration, and a number of business casualties; this is the world of the TV and radio manufacturers—a mature, dangerous jungle. Other parts of the industry are so 'young' that serious worries about competitors' costs are years away; this is the area of electronics where small businesses can still shoot up overnight. . . . Finally, there is that area of electronics—about half the industry—whose affairs are regulated by the clock of history itself: so long as relations between the U.S.S.R. and the U.S. are bad, military electronics will be a good business."

In 1954, when the federal government took its last Census of Manufactures, most electronic products fell into the three industries entitled "Radios and Related Products," "Electronic Tubes," and "Scientific Instruments." * Table 1 gives the employment in those three groups for both the United States and the New York area.

The Census Bureau has already acknowledged the inadequacy of this classification for analytical purposes and has announced a revised grouping for the next Census of Manufactures. The chief culprit is the industry called "Radios and Related Products," which ranges all over the lot—from radio and television sets for the home to radar and navigational aids for the armed forces, and practically all components except tubes and transistors. The variety of the products is accompanied by variety in methods of production and the nature of demand—two aspects which are deeply important in understanding location. Thus we cannot hope to explain geographical shifts in that industry if we treat it as a unit.

In this study, therefore, we shall adopt a classification somewhat more in line with the new plan of the Census Bureau. Our arrangement distinguishes three kinds of electronic products.

1. *Consumer electronics*. This category includes the items with

* Later, in projecting the employment figures for electronics, we will have to add a fourth industry, "Computing and Related Machines." But in 1954 this industry did not contain enough electronic products to justify including it.

Table 1 Employment in the Three Major Census Industries Producing Electronic Items, United States and New York Standard Metropolitan Area, 1954

	United States (thousands)	New York area [a] (thousands)	Area's share of U.S. employment
All three industries	410.3	72.8	17.7%
Radios and related products	294.0	43.8	14.9
Electronic tubes	71.0	11.5	16.2
Scientific instruments	45.3	17.5	38.6

Note: Not all products of these industries are electronic; and some electronic products are found in other industries.

[a] Standard Metropolitan Area of 17 counties, as used by Census Bureau. We estimate that the 22-county New York Metropolitan Region had about 76,000 employees in the three industries.

Source: U.S. *1954 Census of Manufactures,* including unpublished special tabulations.

which the average person is most familiar—radio sets, television sets, phonographs and "hi-fi."

2. *Components.* This category includes the electron tubes and other parts that are made separately and sold to the manufacturers of consumer electronics and also to the manufacturers of items in the third category which follows.

3. *Military and industrial electronics.* This is a convenient label for all electronic products not in the first two categories. A few examples: computers, radar, missile-guidance systems, electronic equipment for planes and ships, commercial broadcasting equipment, two-way radios for police cars and taxicabs, electronic eyes, electronic microscopes, X-ray machines, and automation equipment.

Though employment figures are not yet available for these three categories, reasonably good national figures exist for the dollar value of shipments in 1957, and these are presented in Table 2. Each category will deserve a separate chapter, but not until we examine the locational forces that were at work in the early history of electronics.

Table 2 Factory Sales in Major Categories of the Electronics Industry, United States, 1957 (millions of dollars)

Electronics industry, total		7,600 [a]
Consumer electronics, total		1,545
Television sets		833
Radio sets		374
Phonographs and hi-fi		338
Components, total		2,435 [a]
Resistors		180
Capacitors		225
Transformers		100
Electronic tubes		970
Receiving tubes	384	
Picture tubes	183	
All other tubes	403	
Transistors		70
Other components		890

Military electronics, total	3,900 [b]
Aircraft	1,083
Missiles	1,108
Communication	880
Research, development	303
Ships and harbor craft, combat vehicles, support vehicles, and miscellaneous	132
Industrial electronics, total	1,255
Electronic computers	130
Broadcast, mobile radio, microwave relay, marine and aviation equipment	300
X-ray equipment	100
Laboratory and service equipment	190
Other industrial electronics [c]	535

[a] Only $900 million of components—those used for replacement—are included in the grand total of $7,600 million because the other components are included in the sub-totals for consumer, military, and industrial electronics. This table does not cover components for use in purely electrical circuitry (non-electronic). Tape recorders, though electronic, are not covered because data were lacking.

[b] Estimated by Electronic Industries Association for calendar year 1957. However, the other figures in the military column are for fiscal year 1957 and therefore do not add to the total.

[c] Includes closed circuit television, non-hi-fi sound equipment, industrial controls, radiation and electronic heating equipment, and miscellaneous.

Source: Based on tables in *1958 Electronics Industry Fact Book* (Electronic Industries Association, Washington, D.C., 1958).

THE INFLUENCE OF EDISON

Despite the modernity of electronics, its beginnings are traceable to the times when even the electrical industry was young. Thomas A. Edison was the father of much that is both electrical and electronic. Like Samuel F. B. Morse, the painter who had listened to scientific lectures at Columbia University and had built his telegraph instruments at New York University, Edison found the New York area a congenial place to work.

Edison was at first a telegrapher in the Middle West. He moved to Boston, invented a device to improve the telegraph, and in 1869 journeyed to New York seeking both a market for his invention and a source of capital with which to develop it. He found both. New York's position as the nation's financial center is a part of the Edison story, for he found work in a financial reporting company, branched off on his own, and invented an improved stock ticker which he sold for $40,000. He rented a shop in Manhattan to manufacture the tickers and other telegraphic devices. What happened next has been repeated countless times in the history of industry in New York City: the shop soon became too small.[5]

The move Edison made to Newark in 1870 could be said to have set the pattern—a move from the center of the area to the outskirts. In Newark he started manufacturing in a building at the heart of the city. He quickly opened two more shops and from then on was a large-scale manufacturer as well as an inventor. During the 1870's and 1880's he used various parts of the metropolitan area for experimenting, for demonstrating his inventions, and for financing their manufacture. He invented the phonograph and the incandescent lamp at Menlo Park, New Jersey; proved the feasibility of incandescent street systems by dramatically lighting a portion of lower Manhattan; and opened a lamp plant in Harrison, New Jersey, adjacent to Newark. In 1889 his lighting enterprises were consolidated into the Edison General Electric Company and a search for a suitable base of operations ended in an unfinished locomotive works in Schenectady, far from the New York Metropolitan Region.[6]

The Edison firm had a dominant position in incandescent lighting, though both Westinghouse of Pittsburgh and Thomson-Houston of Lynn, Massachusetts, were also making contributions in the field and had portions of the business. The firm later evolved into the General Electric Company. It is significant that the firm's central office remained in Manhattan and is still there.

The Edison story, in historical perspective, demonstrates some of the kinds of locational events which abound in the later history of electronics. In earlier years, when the Edison operations were new, small, and uncertain, long-term commitments based on a given state of technology were not justified. With limited markets, it was desirable to take advantage of available services and suppliers of the kind found in large metropolitan areas. But with the Manhattan street-lighting project a success, expanding markets were foreseen, both in terms of geography and volume, and Edison General Electric could afford to make long-term commitments and seek advantages of a different kind, such as plenty of elbow room and large facilities capable of catering to a national market.

Edison, while operating in the New York Region, affected the Region not only directly through his work and his enterprises, but also indirectly by his very presence, which attracted others, including Nikola Tesla, A. E. Kennelly, and Reginald Fessenden. Edison's personal attraction is an example of a locational advantage which the Region still possesses: the desire of researchers and engineers to be near the leaders in their fields.

In 1883, Edison noticed something strange going on inside his incandescent lamps. He was then using carbon as a filament; and, as in present-day light bulbs, he was removing the air from the bulb in order to keep the glowing filament from being consumed. He noticed that the inside of the glass became dark after a time. He experimented, and found that when he fastened a metal plate near the filament he could cause a current to pass through the plate. What was happening was that electrons jumped through the airless space to the plate whenever he gave it a "positive charge"—that is, drew some of its electrons away. Not until 1897 would J. J. Thomson of England actually discover the electron; but Edison, in observing what became known as the "Edison effect," built a bridge between "electrical" and "electronic." No man at that time, however, was ready to cross the bridge.[7]

Communication without wires

Signaling through space by means of radiomagnetic waves was made possible by a series of nineteenth-century discoveries, mainly by Europeans. For our purposes, the year 1901 is a convenient starting point. It was then that Guglielmo Marconi succeeded in sending dots and dashes from England to Newfoundland. The transmission was electrical rather than electronic, since he used the electric spark to generate his waves rather than the electron flow later associated with the vacuum tube. Marconi companies sprang up around the world, and the American Marconi Company planted its principal station in New Brunswick, New Jersey, within the present New York Metropolitan Region. Obviously the greatest traffic for wireless would be with the European continent; so it was reasonable that the nation's first wireless stations, both sending and receiving, were along the eastern seaboard, and particularly around the New York area where the concentration of business and of population could be expected to create the greatest volume of business. The electro-physicist, Michael Pupin, was engaged as a consultant by American Marconi while he was a professor at Columbia University, which had established a department of electrical engineering in 1889. Here was an early incident in a trend which has assumed major proportions today, the establishment of professional and commercial ties between universities and business firms.

When Edison had exhibited his curious "Edison effect," one of the interested observers had been an English visitor, Ambrose Fleming. Later Fleming worked for Marconi in England. He believed that a tube like Edison's—having a hot filament and a nearby metal plate—could be used as a detector of aerial signals. In 1904 he patented a device that became famous as the "Fleming valve." With it, he translated oscillating wireless waves into a one-way current by allowing the waves to affect the plate in the tube, so that the plate attracted free electrons from the filament in a pattern conforming to the aerial signal.[8]

Almost at once an American, Lee de Forest, improved the Fleming valve by introducing between the filament and the plate a third object called the *grid,* an arrangement of wire by which he could more sensitively affect the flow of electrons. De Forest patented his tube in 1906. This event really marked the beginning of modern electronics, though some time elapsed before the tubes could be successfully manufactured on a sizeable scale. De Forest had moved from California to New York for financial backing, and, receiving it, had begun his first company in 1901 in Jersey City, in an old rented machine shop. Soon after his modification of the Fleming valve the firm collapsed, and in 1911 de Forest returned to California. He is quoted as saying: "I wouldn't take New York if they offered it to me with a fence around it. I'm going back to California just as fast as I can go; that's the place to live." [9] Even the flexibilities of a metropolitan location did not spell financial success.

While de Forest was operating in Jersey City, Reginald Fessenden, who had once worked with Edison, set up a wireless station there, and another in Brooklyn. His aim was to compete with Marconi. He used a method which later proved to be better than Marconi's, and by his inventions between 1905 and 1913 became a leading figure in the development of radiotelephony. As early as Christmas Eve 1906 he had proved to the world that it was feasible to broadcast music and speech.[10]

But neither Fessenden nor de Forest had the attributes of successful entrepreneurs, and the large-scale development of the wonders of electronics would be left to more solidly financed organizations. Around 1910, three organizations were already in strong positions.

The Marconi interests, operating in the New York Metropolitan Region, were fairly well in control of wireless. The Edison interests, which by then had become General Electric, operating both inside and outside the Region, were dominant in lighting, phonographs, and other electrical developments. General Electric, plunging energetically into research, had purchased a Yonkers plant in order to acquire the services of the German immigrant, Charles Steinmetz. It had also hired Irving Langmuir, who had been born in Brooklyn and studied electricity at Columbia University.[11] Developments in telephony were largely in the hands of a third entity, the American Telephone and Telegraph Company (A. T. & T.), which had moved from Boston to New York in 1907 and was taking a keen interest in the electron tube.

Meanwhile wireless communication—by dots and dashes—was spreading around the world. The first important demand for wireless equipment in the United States came from the government, in particular the Navy, and from steamship companies. The *Titanic* disaster of 1912 demonstrated dramatically the value of signaling without wires. The approach of the first world war, and then the war itself, created a tremendous incentive to both invention and manufacturing.[12] One of the milestones of the decade was reached when A. T. & T. transmitted the human voice from Arlington, Virginia, to Paris. This company also bought the commercial rights to Lee de Forest's many patents. General Electric, during the war period, developed equipment which could be used as the basis for a complete transmitting system.[13] Another domestic firm, Westinghouse of Pittsburgh, manufactured wireless apparatus during the war, and in 1920 gained its own patent position through strategic purchases of inventions by Fessenden and by Edwin Armstrong of Columbia University. But no one company could, on the basis of its own patents, produce the equipment needed for modern radio.

Despite the advances made in the laboratories of domestically owned firms, the greatest source of wireless equipment in the United States during the period 1910–1920 was the Marconi factories in Aldene and Roselle Park, New Jersey. Thus both international wireless communications and the manufacture of wireless equipment remained to a large extent under foreign control. This situation stirred the United States government to urge the creation of an American company endowed with the know-how and the strength to take a leading role in communications. General Electric was approached, and Owen D. Young's organizational innovation, the Radio Corporation of America, was the result.

The birth of this firm was probably the single most important factor in shaping the structure of the electronics industry. RCA was organized at New York City in 1919, and absorbed the assets of American Marconi, which chose not to continue operations in the face of determined opposition. The new firm was conceived as principally concerned with wireless communication, but within a couple of years it was serving as a patent pool for the many communication patents of three vigorous companies, General Electric, A. T. & T., and Westinghouse. Thus RCA, by its agreements with those firms, obtained most of the important radio patents of that day. It also obtained the key station at New Brunswick, New Jersey, and the French station at Tuckertown, New Jersey, thereby acquiring control of the important privately owned stations on the East Coast.[14] It established research facilities on Long Island. It also became the sales organization for products of the "member" firms. Not until much later did RCA itself begin manufacturing.

This study is concerned with the manufacture of electronic products, not the sending of commercial or military messages nor the broadcasting of programs. But the manufacturing industry could not become big-time until broadcasting had created a demand for equipment. The year 1920 brought a springboard for this demand. Even earlier, many enthusiasts had assembled components from here and there and had built sending and receiving sets in their homes. These amateurs were chatting with one another, tuning in on the jumble of voices and code signals in the ether, and having a wonderful time. It was largely to them that the Westinghouse station, KDKA, began regular broadcasts in November 1920 by reporting the results of the Harding-Cox presidential contest. The nation blinked and took notice.[15]

ELECTRONICS COMES OF AGE

During the next few years the manufacture of home radios spread rapidly and wildly, and the American electronics industry, that is, the "consumer electronics" part, was on its way. The New York Metropolitan Region can be reasonably considered to have been the scene of its birth. The Region's great population, its wealth, and its location on the seacoast facing Europe had all contributed to the historic beginnings, but as the industry grew, it grew not only in the New York Region but also in other metropolitan centers, including Chicago, Boston, and Philadelphia.

For about twenty years the electronics industry was mainly the radio industry, and by the time the United States entered World War II, factories were pouring out more than 13,000,000 sets a year.[16] Meanwhile, as before, ingenious men in laboratories were building the shapes of things to come. Television became an experimental reality even before the 1930's, and as early as July 1931 a trade paper, *Electronics,* said in an editorial, "If brought along at a good clip now, it may be the looked-for new industry to bring the army of the unemployed out of the trenches by another Christmas." But when the time for television's commercial advent really came, World War II was approaching, and television had to wait, because electronics was drafted for military service.

The development of television extended over several decades,[17] but commercial exploitation was made possible principally by the work of A. T. & T., Westinghouse, and RCA between 1920 and 1945. Probably the single most important invention, that of a photoelectric tube for television transmittal, was made by Vladimir Zworykin while employed by Westinghouse in 1928. In the period after 1930 the principal burden of research fell to RCA, as has been the case more recently in color television.[18]

The story of invention in the broad area we call electronics does not by any means end with the perfection of television. But the rest of the story is of a somewhat different nature. The work of inventors like de Forest and Edison, carried on in the New York Metropolitan Region where there was a concentrated market in which to test, manufacture, and sell their inventions, helped to establish the Region as of prime importance in the industry. More recently, and especially in the last two decades, though individuals still have been recognizable in their contributions, an increasing proportion of the developmental work has been carried on within governmental installations, large firms, and university research laboratories. Moreover the great proliferation of electronic products has made the contributions of individual inventors harder to trace. The mounting complexity and cost of research have encouraged its institutionalization, and so have the requirements of national security and military procurement.

In World War II, the most dramatic military application was radar, but electron tubes were also used increasingly in weapons, "walkie-talkies" and other military communications systems. The United States government, which had spent only $100 million on military electronics in the year ending June 30, 1941, spent an average of $1.5 billion a year during the next five years.[19]

Civilian radio production stopped during the war, and when peace was restored the pent-up demand was huge. Twenty million radio sets—a new record—rolled off the assembly lines in 1947. But that was the best year radio has ever had. Television now arrived at last, and its fabulous three-year boom period began. In 1947 the industry produced 179,000 television sets. In 1950 it produced nearly 7,500,000, a total that was surpassed only once in the next seven years.[20]

As the 1950's began, consumer products were an even larger proportion of all electronic items than they had been in 1939. But then came the Korean War and the world-wide Communist threat, and the government moved back into the field, boosting its expenditures for military electronics from $500 million in the calendar year 1950 to $3.9 billion in 1957. During the same years, what we have called "industrial" electronics, including computers, data-processing machines, and new laboratory instruments, began to hit its stride. In 1950, "industrial" electronics was only one-fourth the size of consumer electronics, in the dollar value of products, but by 1957 it had almost caught up with consumer electronics.[21]

By the late 1950's, then, the three main branches of the electronics industry—consumer items, components, and military and industrial equipment—were in different stages of development. Each differed from the others in its locational needs. And in each the New York Metropolitan Region played a somewhat different part.

2

Electronics in the Living Room

In the big radio craze of the early and middle 1920's, the New York Metropolitan Region was the nation's chief center for the production of radio equipment for the American consumer. By 1929 the Chicago area had surpassed it, and Chicago has kept the lead ever since.[1] When television swept the country in the late 1940's, the New York Region, though never *the* center as it had been in radio, possessed at first a larger share of TV manufacturing than it does today. Like radio, the TV manufacturing scene is now dominated by factories in the Midwest.

In short, the Midwest has become the leading section in consumer electronics. Table 3 shows how far it led the Middle Atlantic states in 1954 in the production of TV sets, radio sets, and phonographs. It shows, too, that Illinois alone had a greater volume of factory sales of those products than the Middle Atlantic section, which contains almost all of the New York Metropolitan Region and also the important Philadelphia-Camden area and the large plants in upstate New York.

Most radio-TV manufacturers have two major characteristics in common. First, their market is mainly the American household, rather than other manufacturers.* Second, most radio-TV manufacturing now takes place in large plants. Some assembly lines produce more than a thousand sets a day. The component parts are

* The main exception is the sale of radios to the automobile industry; this is an important exception because more than one-third of all radios—more than five million a year—now go into cars, but it is not an exception that favors New York, given the concentration of auto plants in the Midwest.

Table 3 Geographical Distribution of Television, Radio, and Phonograph Manufacturing, by Value of Shipments, 1954

	Total, TV receivers, radios, and phonographs	TV receivers, including TV-phonograph combinations	Radios, phonographs, and radio-phonographs
United States [a]	100.0%	100.0%	100.0%
North Central [b]	59.5	57.7	64.6
Illinois	34.7	33.6	37.9
Middle Atlantic [c]	33.4	35.7	26.9
New York State	15.1	14.5	16.5
South	3.7	3.3	4.7
New England	0.8	....	2.9
West	2.6	3.2 [d]	0.8

[a] U.S. total was $1,384,560,000, consisting of $1,020,992,000 television sets and $363,568,000 radios and phonographs. Because of rounding, percentages do not always add to exactly 100.

[b] Consists of 12 states, but nearly all of the manufacture of these products was in Illinois, Indiana, and Ohio.

[c] New York State, New Jersey, and Pennsylvania.

[d] All in California.

Source: U.S. *1954 Census of Manufactures.*

fed in continually from outside suppliers—or, in a few cases, from other facilities of the same firm.

Both the distribution of the market and the size of the production plants have changed since the early years of radio manufacturing. And so has the structure of the industry. These changes have had great significance for the New York Metropolitan Region.

Evolution of radio manufacturing

Many circumstances favored radio manufacturing in New York in the early 1920's. We have already encountered the circumstance that the Region was the birthplace of much of the technology. Indeed, it has continued to be a great center for experimentation, invention, and technical development, to the present day. We also saw in the first chapter that the Region was producing wireless equipment before radio broadcasting began.

Another important circumstance was the early rise of New York's local market for home radios. The Region's great population would have made it an important market in any case, but it also had a headstart in manufacturing because it became the principal center for broadcasting. Networks did not enter the picture until 1926, and as long as broadcasting had to be conducted on a local basis the New York area with its array of stations and entertainers offered special advantages to listeners. Thus a big demand for radios appeared there before it appeared in most other places.

Finally, and perhaps of greatest importance for our understanding of later developments, the typical producer in those years ran a very small enterprise.

When new products are born, the technology of production does not usually demand large capital investment. It is only later, when the technology settles down and when specialized machinery and specialized plant layouts come into use, that the need for a large capital investment may develop. There are exceptions, of course, as witness the production of nylon. But the exceptions aside, early conditions in many industries have permitted the entry of small firms with minimum capital resources.

Radio was such an industry. In its early years it consisted of numerous small establishments. Demand grew swiftly; the public gobbled up products without discriminating sharply; and many entrepreneurs, including some who found rocky going in other businesses during the 1920–1921 recession, rushed to get in on the act, as they could still do without investing much capital. Yet uncertainty loomed over them all. No one could predict how fast the aggregate demand would grow, and no firm could predict what its own share would be. What is more, methods of production were untested. The product itself was being modified rapidly and was not yet predictable from year to year. The uncertainty was heightened by the greater seasonality of demand in those years, because radio reception was then rather poor in the summer months.

Thus not many borrowers nor lenders cared to make a huge investment in plant. Instead, commitments were limited until the initial enthusiasm would subside and a more reliable assessment could be made of the future prospects of the industry. At the same time competitors had to be closely watched, if not imitated, in order to incorporate the latest developments into the product.

Under all these conditions a metropolitan location was convenient, and the New York Metropolitan Region was particularly attractive to the producer. Here he could alter his product or his volume of output with relatively little strain on finances because most of his costs could be kept variable. Quarters of various sizes and suitability could be rented; a large pool of labor accustomed to factory procedure was available; and suppliers close at hand were ready to fill needs which varied both in specification and in quality. Here the producer could keep abreast of the latest developments in the radio arts.

Though the large existing electrical companies interested themselves in radio, they were far from equipped to absorb the tidal wave of demand. In 1921, there were said to be about two dozen small companies in the United States making "parts" for radio enthusiasts to put together,[2] and the number quickly ran into the hundreds. The best way for a manufacturer to attain large-scale production and thereby to cut costs was to assemble parts into complete sets, which could be sold under a brand name as an easily recognizable piece of furniture instead of a gadget. Radio would thus come into the living room from its hideaways in the attic or basement. Westinghouse began making complete sets in 1921, and General Electric in 1923. But numerous manufacturers in those early years, because of the state of innovation, uncertainty, and chaos in the young industry, were at first unable or unwilling to invest in facilities to make complete sets. At the outset of 1923 there were about a million and a half sets in operation, but about 75 per cent of these had been put together by the public.[3]

One can find a close parallel in the progress of "high fidelity" in the 1950's. Until recently all hi-fi fans bought an assortment of gadgets, made by various manufacturers who specialize in things like tuners, amplifiers, turntables, or speakers. The popularity of such equipment has grown so much, however, that the consumer can now find on the market "packaged" high fidelity, that is, components chosen by the manufacturer and assembled into a cabinet to be sold as a unit. The purchaser no longer has to know the difference between a "woofer" and a "tweeter" to own a set which he can show to his neighbors as hi-fi. Radio of the twenties went through this same evolution.

In the years from 1923 through 1926, complete radio sets gained in popularity, and their manufacture lured additional hordes of newcomers. As shown in Chart 1, some 750 radio manufacturers are estimated to have entered the industry in the four years be-

Chart 1

Entry and Survival of Firms Manufacturing Radio Sets, United States, 1923–1934

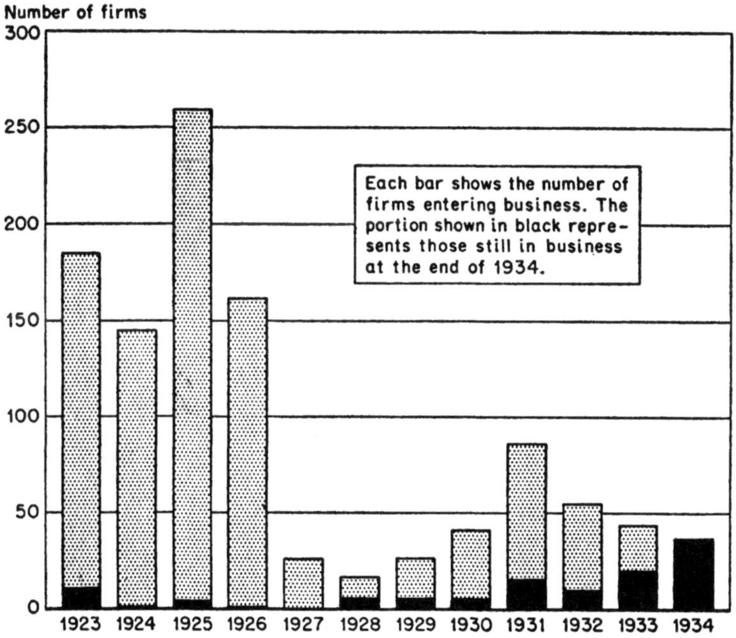

Source: *Electronics* Magazine, March 1935.

ginning with 1923. Many of them abandoned the field of battle as suddenly as they had arrived. Only about 20 or 30 of these competitors were still in business in 1934—but that takes us ahead of our story.

The available statistics of the twenties are ragged, conflicting, and overlapping. By 1925, it is asserted, something like 3,000 manufacturers were engaged in some aspect of radio manufacturing.[4] Perhaps 400 of these were turning out complete radio sets.[5] The rest apparently were making parts, either for the set-manufacturers or for consumers who still preferred to assemble their own radios. Unfortunately we have no statistics worth giving on the location of the 3,000 firms (if that was the actual number). Nor can we geographically divide the perhaps 400 set-manufacturers, and we do not know how many of the workers they employed were in the New York Metropolitan Region.

We do know that the "3,000" and the "400" were in both big and little cities throughout a belt extending from the East Coast to Chicago, and that a large number of them were in the New York Metropolitan Region, on Manhattan's Varick Street and in places like Brooklyn and nearby New Jersey. A few of the New York plants were sizeable. The Charles Freshman Company expanded its capacity from 7,500 sets per week to 16,500 between 1925 and 1926, besides making kits for home assembly and parts. The Freed-Eisemann Radio Corporation, operating in Brooklyn plants finally totaling 200,000 square feet, went from a sales volume of $250,000 in 1922 to over $5 million in 1925.

Still another factor affecting the size of radio plants was the question of patents. The Radio Corporation of America held the patent rights on many radio inventions which had been developed or acquired by General Electric, Westinghouse, and A. T. & T. Firms exploiting such inventions without a license from RCA were open to lawsuits, but RCA was ready neither to satisfy the sudden consumer demand nor to police the infringement of its patents as hundreds of imitative small manufacturers swarmed into the industry. There was much confusion about the validity of patents, but this seems clear: the fear of possible prosecution acted as a further incentive for manufacturers to be small and obscure.

In the late 1920's, however, the radio industry began to take on a completely different structure. The multitude of small firms melted away, leaving a reduced number of large firms, producing complete sets and devoting considerable effort to building up consumer loyalty to brand names. Probably the most potent force bringing about this development was RCA's growing enforcement of its patent rights. Aided by a series of court actions, RCA fashioned a licensing policy that virtually compelled the licensee to operate on a large scale in order to survive. In fact, for several years every patentee was required to pay annual royalties of not less than $100,000.[6]

But there were other important forces impelling the industry toward increasing concentration. One was the slowing down of the changes in radio technology. It became safer to make large commitments in plant facilities, for the investment could be recouped in reduced manufacturing costs. The product could now be made more standardized. This meant that production could be more repetitive, that large plants could be built at locations apart from the great metropolitan centers, and that unskilled labor could be recruited at the lower wage rates typically prevailing in those smaller labor markets.

Besides, the consumer, his initial hunger for radio having been satisfied, had come to suspect the integrity of the small, uncertain producer who could not afford to guarantee the quality of his product. Edison himself, nearing the end of his long life, was quoted as saying in 1926: "The radio is a commercial failure, and its popularity with the public is waning. Radio is impractical commercially and esthetically distorted, and is losing its grip rapidly in the market and in the home."[7] Responsible firms saw the need to recapture public confidence. Nation-wide advertising, with its cost spread over a large volume of output, was seen as the answer. Styling developments, as a stimulation to sales, came to be almost as important as technical improvements. The proliferation of radio and television models today is an extension of this early shift in emphasis.

By 1927 the number of set-manufacturers had declined precipitously. As the small firms had faded out by the hundreds, not many had come in to replace them, as shown in Chart 2. The competition was too formidable; the entrepreneur who dared to invest in radio facilities found that the sellers' market of earlier years had turned into a buyers' market. Even sizeable firms felt the severe squeeze

Chart 2

Number of Set-Manufacturers Entering and Leaving Radio Industry, United States, 1924–1934

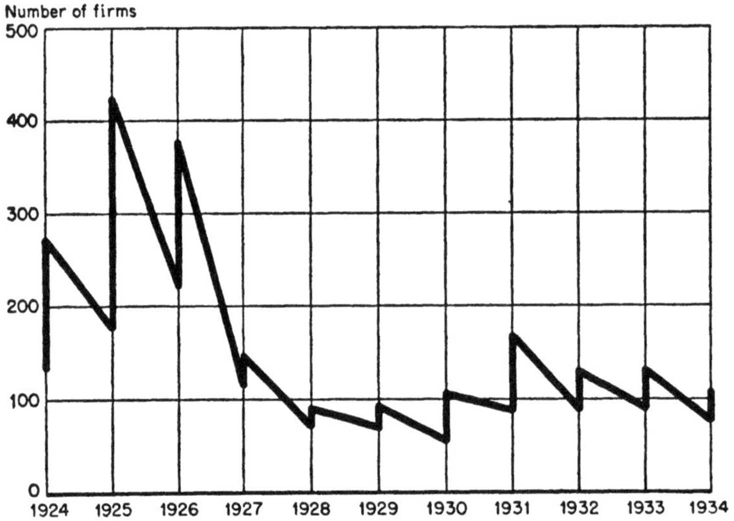

Each vertical line represents the newcomers for the year indicated, and the oblique line immediately following shows, by the extent of its drop, the firms that went out of business in that same year. The high point thus shows the number of firms in business during some part of the year.

Source: *Electronics* Magazine, March 1935.

to which the industry was subjected. The performance of the two New York firms mentioned above may help to illustrate the force of the squeeze. The sales of the Charles Freshman Company in each of the years 1925, 1926, and 1927 were slightly over $7 million. Net income, however, dropped from $1.4 million to $700,000 and then, in 1927, to a loss of almost half a million dollars. The stock of the Freed-Eisemann Radio Corporation sold as high as $33 in 1925, as low as $1.50 in 1927. In 1928 the two firms merged and removed all manufacturing operations to Clifton, New Jersey. In that same year $2.5 million was lost.[8]

In 1930 and 1931 the industry received a temporary lift through the introduction of the "midget" set. Eighty new firms entered the field. But as the depression deepened and the price of radios fell, the trend towards concentration gained renewed vigor. Over-production added to the industry's woes.[9] It was estimated that "radio is carrying the load of a factory capacity 10 or 12 times any production the trade can possibly absorb."[10] Profits were nonexistent except for diversified companies, and only the strong survived.[11]

Just as the structure of the industry was greatly influenced by the creation of RCA and its agreements with other firms in the radio group, so the dissolution of the radio group in 1932 foreshadowed further changes. Under a consent decree in an anti-trust case, RCA was to establish itself as a manufacturer while General Electric and Westinghouse were to refrain from manufacturing and selling radio apparatus for two and a half years. This development fostered the growth of other firms which are today major factors in the industry both in radio and television. Companies such as Philco, Zenith, Emerson, and Galvin (Motorola) capitalized on the RCA consent decree.[12] Both GE and Westinghouse subsequently re-entered the market and have been joined by more recent entries such as Sylvania and Admiral. In radio manufacturing today, though specific figures are unavailable, about ten firms account for the overwhelming volume of sales—conservatively estimated at about 80 per cent.

Of these ten, the only one which has survived in the New York

Metropolitan Region through the whole period of change is Emerson Radio and Phonograph Company. It now makes both radio and TV sets in the Region. Westinghouse, departing from the pattern typical of the industry as a whole, entered the Region in 1951, opening a large radio-TV plant at Metuchen, New Jersey, near other Westinghouse plants in that area.

The shift toward large-scale radio production that began in the late 1920's coincided with the extension of radio broadcasting throughout the nation. The NBC and CBS networks were formed in 1926 and 1927. The demand for receivers, which had been concentrated in large centers and mostly in the eastern part of the nation, was spreading through the countryside.

Thus, while the original ties to the New York Metropolitan Region on the production side were being weakened by the transformation of the industry, the market attraction of the Region's multitude of consumers was also being dissipated with the emergence of a national market. Under this new set of conditions the Region did not fare very well relative to national production. Many of its principal advantages having been undermined, its disadvantages grew in importance. We need to consider these disadvantages of a location in the New York Metropolitan Region, but first we must extend the boundaries of our problem to include television.

Television—where and why

In the boom years of TV manufacturing, starting with 1947, many new firms of modest size blossomed in the New York Metropolitan Region for the same reasons that radio firms had appeared in the early 1920's: the best television programs were there; small manufacturers found advantages in the environment there; and the demand was so heavy that the public was not insisting on the brands of well-known firms.

On the other hand, television broadcasts became widely available much more quickly than had radio broadcasts. Besides, the technology of television settled down very fast, and large-scale production of standardized products did not have to wait. The result was that the established radio manufacturers were the ones best prepared to launch this new industry. They had the engineering know-how. They possessed a reservoir of good will built up around their brand names. They had the financial power to make the necessary investment in equipment and inventories; and such investments were greater in television than they had been in radio, so great, in fact, that small-scale production was far more hazardous than it had been in the 1920's. These strong radio firms, as we saw, had already found that locations outside the New York Metropolitan Region were more suitable for their needs, and there was nothing in the manufacture of television sets that demanded a different part of the country.

Once the initial demand had been satisfied, there was in television, as in radio, a dramatic reduction in the number of manufacturers. From a peak of about 140 set-makers, the number fell to approximately 50 in 1956. Numerous small firms in the New York area were among those that went under, and one large national advertiser discontinued its manufacturing of TV receivers in Long Island City.

It is estimated that in 1957, some 85 per cent of all TV sets were being produced by ten manufacturers—a situation closely similar to that in radio. An informed guess would rank them about like this: RCA, GE-Hotpoint, Zenith, Philco, Admiral, Motorola, Silvertone (Sears, Roebuck), Emerson, Magnavox, and Westinghouse. Exact figures are impossible to obtain since the firms regard their share of the market as an important trade secret. Of this group, only Admiral was not in the forefront of radio.

As already pointed out, Emerson and Westinghouse have their plants in the New York Metropolitan Region. GE and Philco are also in the East, at Syracuse and Philadelphia respectively. The remaining six manufacture their television sets in the Midwest, and four of them, Zenith, Admiral, Motorola, and Silvertone, are in the general vicinity of Chicago.

Thus, while the New York Metropolitan Region did not have a

very large share of the TV industry to begin with, its position declined still further in the shakeout which followed the stabilization of demand and production.

THE STRENGTH OF THE MIDWEST

So we see that radio-television manufacturing, having achieved maturity, has evolved a pattern of location quite different from that which accompanied its early development. Why have other locations been preferred over the New York Metropolitan Region?

Ordinarily, in analyzing the locational preferences of industries, one finds an important guide in the cost structure. Only a few firms make most of the component parts of their radio and television sets; for the others, the production process is one of assembly. It is possible to give a range of cost figures within which the great majority of establishments will be found. For example, executives interviewed by the author estimated that a hypothetical television set costing $100 on the shipping dock will require a direct labor cost running between $7 and $12, with the average being about $9.[13] The cost of materials on the same set would range between $65 and $82, the typical cost being about $78. About $20 of this material cost is in the picture tube. Tooling and engineering costs may run to $3. The remainder is in other items such as selling, administration, and advertising, which can be lumped together as overhead, totaling about $10.

Materials. The cost of materials, being such a large proportion of total costs, might appear at first glance to be the most critical factor influencing the location of radio and television plants. But this is not so, for the cost of materials does not differ crucially as between one location and another in the broad band which includes both the Northeast and the Midwest. That is the conclusion which emerges from our interviews with manufacturers both inside and outside the New York Metropolitan Region.

In general, radio and television sets are made from a variety of parts that are widely produced and cost little to ship. The bulkiest items are cabinets and TV picture tubes. As for picture tubes, they are produced in large volume both in the Middle Atlantic and midwestern states. True, the other heavy item, the radio or TV cabinet, once exerted a locational pull to the Midwest. A supply of hardwoods suitable for veneers and the presence of a furniture manufacturing center at Grand Rapids contributed to the pull. This attraction to the Midwest has now greatly diminished. Many of the lower-priced sets now have metal or plastic cabinets, and, besides, the South has become the principal source of wood for cabinets. The Magnavox plant at Greeneville, Tennessee, is well placed in relation to hardwoods from western North Carolina.

Thus it is not likely that the pattern of location for the manufacture of sets is determined significantly by the desire to economize on incoming freight costs, though the need for wood cabinets was at one time a considerable factor.

Labor. We have already seen that in the production of a television or radio set the cost of labor is small compared with the cost of materials. Even so, labor could have locational significance if its supply or its price differed sufficiently between one area and another. Let us see whether this is so.

The large nation-wide demand for radio and TV sets and the standardization of the product have made it possible for manufacturers to seek the economies of large-scale production. These economies have been achieved by breaking the production process into a great many small operations. For example, there are more than 1,600 separate operations in the assembly of the 600-odd parts of a television set. The work is divided with such extreme specialization that most of the operations are performed within a 30-second cycle.

This method of production requires a large number of workers in a single establishment, generally from 1,000 up. It also requires a high proportion of unskilled or semiskilled workers. Most of the people in radio-TV plants are women. Engineers, draftsmen, tool-and-die makers, and machinists are not necessary in such numbers as to constitute a severe limitation on where the plant is located.[14]

A radio-TV manufacturer gets along on a much smaller proportion of skilled employees than, say, a firm on Long Island fashioning military electronic products.

Thus a radio or television plant, despite the big labor force that it requires, does not have to be located in a huge metropolitan area. Medium-sized cities will do, especially in areas where surrounding agricultural districts are constantly releasing surplus labor. Indeed there come to mind two examples of plants in rather small localities—Sylvania's plant at Batavia, in western New York State, and Magnavox's plant at Greeneville, Tennessee. On the other hand such establishments can find suitable work forces in Chicago, and also in the New York Metropolitan Region.

So far the discussion of labor requirements offers little to explain the locational pattern of radio and television manufacturing. But when we turn to wage rates and to the question of unionization we find clues, at least to the attraction exerted by medium-sized and small cities.

Comparative figures on wages in radio and television plants are scanty because these segments of electronics are not generally separated in federal statistics from other segments. Some rough comparisons can be made, however, with figures gathered by the author from confidential sources. These figures are average hourly earnings in 1957 for a group of occupations representative of television manufacturing. They suggest that earnings in the New York Metropolitan Region were not high in comparison with those of other large metropolitan areas where TV sets are made, but were considerably higher than those of the smaller localities in the industry. The Region's average for these occupations was $1.80 an hour, about on a par with Philadelphia and Chicago. But in some places in Indiana the average was as low as $1.62, and there were places outside Chicago with rates in the neighborhood of $1.50.

Moreover the whole electronics industry is characterized by a variety of competing unions and by disparities not only of wage rates but also of fringe benefits, seniority rules, and occupational differentials. Employers tend to view all these things as being, on the whole, less favorable to them in the largest metropolitan areas than elsewhere. In smaller localities, whether in the Midwest or elsewhere, an employer might well expect to find labor unorganized, or organized by a union of his preference. Many an employer in a large metropolitan center, contemplating his high wage levels, the narrowing of the wage gaps between unskilled and skilled workers, and an aging work force protected by seniority rules, has welcomed an opportunity to start over with a new work force and a revised wage structure.

The degree to which these labor differentials have actually worked against the New York Metropolitan Region in radio and television manufacturing is not easy to measure. Their effects show up more clearly in the making of components, and therefore labor will be examined again in the next chapter. In radio and television assembly, labor must be viewed as a contributing factor in the shift away from the New York Metropolitan Region to the Midwest, but it does not tell the whole story of the midwestern domination. Another explanation is needed.

Serving a national market. So far we have analyzed the location of set-manufacturing in terms of where it is most convenient to carry on the productive process. There is the further question of choosing a location from which it is least costly to serve the market. And here we find a clear-cut preference for a location in the Midwest.

In the simplest terms, one disadvantage of the New York Metropolitan Region is that it is off center in relation to the geographical distribution of purchasing power throughout the nation. If we take personal income as a convenient measure of purchasing power, we find that between 1929 and 1954, the Northeast's share of the national market decreased from 38 to 30 per cent and the share of the Midwest remained constant at 31 per cent, while the South increased from 19 to 24 per cent and the West from 11 to 15 per cent. If we were to assume that on balance New York were at no competitive disadvantage insofar as the southern market is concerned, the greater proximity of midwestern locations to the growing western market would still tip the balance in favor of the Midwest.

Can we translate this disadvantage into dollars and cents? In the first place it should be emphasized that radio and TV sets are sold f.o.b. factory. The distributor must pay the freight, and regional prices are set accordingly. National advertising does not generally name a price, the main exceptions being those cases in which a manufacturer launches a special promotion campaign.

The figures in Table 4 are illustrative of the freight rate disadvantage of the New York Metropolitan Region as compared to Chicago in serving markets in the West and in parts of the South. The actual rates are not shown—only the difference in cents per hundred pounds. The figures can be taken as applicable to a console television set weighing 100 pounds.

Table 4 Differentials in Rail Rates on Television Sets to Selected Cities from New York and Chicago, 1955

Destination	New York's differential lower or higher than Chicago (per 100 pounds)
Tampa, Florida	$ .06 lower
Columbia, South Carolina	.43 lower
Atlanta, Georgia	.37 higher
Dallas, Texas	2.50 higher
Denver, Colorado	2.18 higher

Source: Estimates of freight rate differentials drawn from *Fast Frater* published by the Chicago and Eastern Illinois Railroad, 1955.

This is not to say that producers in the East do not penetrate markets in the Midwest and West and vice versa. In fact a transportation survey conducted as a part of the New York Metropolitan Region Study showed that radio and TV plants in the Region do have national markets. This is shown in Table 5. And it is common

Table 5 Estimated Distribution of Outbound Freight of Radio and Television Manufacturers in the New York Metropolitan Region, by Area of Destination, 1956

Destination		
Middle Atlantic, total		37%
New York Metropolitan Region	25	
Other Middle Atlantic destinations	12	
New England		6
East North Central		16
South		17
Mountain and Plains states		13
Pacific Coast states		8
Rest of world		3
Total		100

Source: Derived from a study of transportation in the New York Metropolitan Region, to be reported by Benjamin Chinitz in a forthcoming volume in this series. The estimates are based on a sample of plants whose total employment in 1956 was 7,000.

knowledge that products of midwestern firms are sold in the East. There is sufficient loyalty to particular brand names to permit this cross-penetration of markets. Thus freight costs are not prohibitive to the point where each market is protected for its local producers. At the same time, the producer aiming at national distribution would still prefer a midwestern location to enhance his competitive position. This preference goes far toward explaining the trend toward location in the Midwest once the industry had settled down to large-scale production.

The trend has also been abetted by the changing structure of transport costs. During the quarter-century between the early days of radio and the postwar television boom, the truck graduated from local cartage to inter-city transport. This has had significant effects on the relative attractiveness of areas as production sites. Trucking made sheer distance a much more important factor and access to rail and water terminals a much less important factor in determining transport costs. The advantage of being located closer to the market was therefore enhanced. The producer not only reaped a reward in lower transport costs but also was able to offer much more rapid delivery. It mattered much more now whether he was 500 or 1,000 miles away from his market.[15]

THE RETURN OF THE PHONOGRAPH

Until the 1920's the hand-crank phonograph, an Edison invention, had a position of honor in the American parlor. The radio took its place. In the 1920's and 1930's the fallen phonograph went electronic—that is, acquired tubes—and after World War II, phonographs came back on the market by the millions, most of them installed in the same cabinets with radio sets.[16] After 1947 the phonograph dipped in popularity while the nation's families were captured and spellbound by television. But then the introduction of long-playing records opened the way for a new phonograph era. In the 1950's a new, full-bodied sound, like nothing that had ever before issued from a parlor gadget, was heard increasingly in the nation's households, and a new word took a firm position in the language: "hi-fi."

High fidelity sound reproduction was not at first performed by complete phonographs, but by hook-ups of sub-assemblies—such as tuners, amplifiers, turntables, speakers—which music-lovers purchased separately and distributed about their dens or living rooms. New York became a center for the manufacture of such articles, and many of them were, and are, imported from abroad.

But as hi-fi increased in popularity, the big electronics manufacturers started "packaging" the necessary components, assembling them in the same kinds of plants that turned out radio and TV sets. By the late 1950's, phonographs, most of them now occupying cabinets of their own rather than being combined with radio or TV, were experiencing a major revival.

The rise of the packaged hi-fi phonograph brought a shift of hi-fi production away from New York, for phonographs, whether hi-fi or not,* are produced in the same general geographical areas as radio and TV sets. As we have seen, the New York Metropolitan Region, though it is one of those areas, is not a dominating one. But the production of hi-fi equipment in the more traditional sense of sub-assemblies, custom installations, and expensive packaged items is still rather heavily concentrated in the New York Region. This kind of equipment is produced largely by a group of manufacturers whose volume of business is not high. Their cost structure is different from that of the radio and TV set-manufacturer. The skill element plays a more important role. Product development carried on by the individual company is more intense. In sub-assemblies, this intensity of research and development pays off, because the customer for such equipment is usually conversant enough with the technology to be impressed with refinements to which the ordinary customer would be oblivious. It is safe to say that the average hi-fi factory is much smaller than its counterpart in radio and television.

The future of hi-fi, including its latest wrinkle, stereophonic sound, is a question to be discussed in the final chapter.

THE CITY VERSUS THE OUTSKIRTS

In consumer electronics, the trend away from the New York Metropolitan Region as a whole has been accompanied by a shift of facilities within the Region from the center to the suburbs. During the early years of radio most of the activity took place in Manhattan. It was there that the producer most readily obtained the degree of flexibility which was so critical to the operation of a small enterprise in a new industry. But as the industry settled down to large-scale production the need for adequate space for the storage of parts—especially in television—and for the proper layout of machines prompted many producers who wanted to remain in the

* There is no generally accepted boundary line between phonographs that are hi-fi and those that are not. Some companies have advertised low-priced "hi-fi" sets that other companies do not recognize as hi-fi.

vicinity to seek new locations outside Manhattan. One or two illustrations will serve.

Olympic began life in 1935 as Namco, a loft operation in Manhattan making radios for the export market. During World War II the firm was enlarged and renamed Hamilton Radio, specializing in electronic equipment for the armed forces. After the war the name Olympic was adopted and the company went into production of civilian radio. The quarters became cramped, the trucking situation became less satisfactory, and the approach of television provided the impetus for a move. In 1948 a multistory factory became available at a favorable figure in Long Island City and the move was made there just in time to take advantage of the television boom.

Emerson, long in the forefront of radio, converted some of its radio assembly lines in 1948 and produced television receivers at a rapidly increasing rate. Unit for unit the manufacture of TV sets required at that time more than ten times as much space as the manufacture of radio sets. To meet these space demands Emerson in 1950 vacated 300,000 square feet of rented space on Eighth Avenue and purchased a five-story structure in Jersey City containing 450,000 square feet of space. The firm later enlarged this plant.

In general, radio and television manufacturers in the New York Metropolitan Region have shown, in recent years, a preference for New Jersey locations over locations east of Manhattan. This preference cannot be explained in terms of lower space costs, the factor which accounts for the trend away from Manhattan. Instead, New Jersey seems to be preferred because of freight advantages. During interviews with industry executives we saw evidence and heard complaints of the difficulty of doing a volume production in the Long Island boroughs of Queens and Brooklyn, particularly if rail shipments were involved. It is hard to find manufacturing space with adequate rail facilities, and even after cars have been loaded there is the problem of getting them to the west side of the Hudson River where they can start the inter-city haul. This introduces additional uncertainty in the shipping schedule which in turn affects the production schedule. Shipping bottlenecks are especially troublesome when the producer must deliver for a special sales event of a distributor or dealer. Although no similar problem exists in the case of trucking, the producer nevertheless prefers the New Jersey side where he can enjoy more reliable rail service as well as a head-start towards the greater part of his market by any mode of transport.

Summary

In this chapter we have examined the forces affecting the location of one major branch of the electronics industry, namely, consumer electronics. We found that the New York Metropolitan Region played a big role in early development by providing an attractive environment for the small producer. When radio and, later, television settled down to large-scale production they grew more rapidly elsewhere—primarily in the Midwest. The New York Metropolitan Region was at a disadvantage because of its geographical position with respect to the national market and because its labor costs were higher than those in smaller cities and towns. The same factors were at work in the manufacture of phonographs, which were enjoying a national comeback in popularity; but the Region remained a prime center for quality hi-fi items produced in plants of modest size. Within the Region, the increasing need for space in the manufacture of consumer electronics caused a shift from Manhattan to locations farther out.

3

The Bits and Pieces

Television and radio sets and all sorts of other electronic end-products, including computers and radar, are constructed from variations of the same basic building blocks. The manufacture of these components is an important industry in itself.

First there are the electron tubes. In radio sets they are receiving tubes, and in TV sets they are both receiving tubes and picture tubes, and these two types together still dominate the field of components. Nevertheless a great upheaval now has begun; it was foreshadowed in 1948 when the Bell Laboratories announced the invention of the transistor, a tiny, durable device which can perform the functions of a tube, with the electrons passing through a solid crystal material instead of through a vacuum. Still other "solid" devices have joined the transistors, and the whole fast-expanding group goes under the name of "semiconductors" because they are neither full conductors of electricity nor insulators.

In addition to the tubes and semiconductors there are many other devices that control electrical currents, items like capacitors, resistors, transformers, filters, and speakers. Finally, though not of much importance for our purposes, there are the "hardware" items such as knobs, dials, and sockets.

Trends in location

A substantial part of components production goes for replacement, as anyone realizes after the demise of his picture tube, but most of the output of component-makers is sold to other manufacturers. Component-makers catered to the electrical industry long before the electron tube made the radio industry possible. As radio became more than an oddity or hobby, these same firms began to supply standard components, which could be used with the radio circuits being developed. In fact, many of them also supplied parts which were sold to the public for home assembly, as mentioned in the preceding chapter. In those early years, the boundary between such products and the "components" sold to other manufacturers for factory assembly was not very clear. With demand growing fast, new firms were formed which directed their major efforts to supplying the new industry. The New York area cradled the components industry just as it did the manufacture of radios, though again Chicago quickly became an important production center, as did Boston and Philadelphia.

At this early stage a relatively high proportion of the cost of making components went for research and development. The lack of well developed manufacturing machinery made skilled workers a necessity. These were to be found in New York and other large cities where the electrical industry had preceded the growth of radio. Moreover, because of the dizzy pace of technological advance, close liaison between set-manufacturers and their suppliers was much more important than it is now. The risks of obsolescence hovered over both the makers of components and the assemblers of receiving sets. Proximity to one another lessened these risks for both, since face-to-face contact facilitated the exchange of information on technological advances, and since inventories could be held to a minimum consonant with satisfactory service to the market. All of these factors accounted for a clustering of makers of components in the metropolitan areas and for the congregating of a large number in New York and vicinity.

In the case of vacuum tubes, the New York Metropolitan Region became of foremost importance. In fact it had been a scene of tube-making even before the arrival of radio broadcasting. For example, the Western Electric Company started manufacturing tubes in 1912 on West Street in Manhattan, and moved in 1925 to Hudson Street where it remained until 1947. But it was the Newark area that came to dominate the Region in tube production. It still does, chiefly because RCA, the largest producer of receiving tubes, has some of its electron-tube plants in adjacent New Jersey towns. Edison's electric lamp industry had originated in the Newark area, and the manufacturing technology learned by workers in the lamp plants was easily transferable to the assembly of tubes.

But tube-making followed lamp-making in other areas of the country besides Newark. These were not all metropolitan areas. Sylvania, one of the giants of tube production, has its principal receiving tube establishment in Emporium, Pennsylvania. Two predecessor companies were established there in 1905 and 1924, devoted to carbon lamps and radio tubes respectively. In the Boston metropolitan area, the early manufacture of lamps laid the foundation upon which Raytheon and Hytron (now CBS-Hytron) were built. Not only the technology of tube-making but also the composition of the work force was historically similar to that of the lamp industry. In both industries the initial models and production runs were handled by male employees with well developed skills. Women began to enter America's labor force in increasing numbers in the period after 1920, which coincided with the growth of large-scale demand for radio tubes. The production of receiving tubes expanded from 4.5 million a year to 69 million in the period 1923 through 1929.[1] The plant operations required manual dexterity and close attention, while at the same time being repetitive and dull. In such work, men are no match for women, as demonstrated many times in industrial history. Today, in some tube plants, women constitute as much as 85 per cent of the work force. As will be explained presently, the manufacture of tubes, that is, of the receiving tubes and picture tubes which are produced in largest volume, has moved increasingly away from metropolitan areas.

A recent tabulation lists 14 makers of receiving tubes.[2] The three largest firms, RCA, Sylvania, and GE, make 70 per cent of all receiving tubes. Others high on the list are Tung-Sol, recently merged with Chatham Electronics; CBS-Hytron, Raytheon, Philco, and Westinghouse. Two of the firms, RCA and Tung-Sol, have their receiving-tube headquarters in the New York Metropolitan Region—at New Jersey locations—and some of the others also produce these tubes in the Region but derive their major income from other tubes of specialized types. Neither Sylvania nor GE produces receiving tubes in the Region.

As for TV picture tubes, the same three firms lead the field. Sylvania, Number 1, and RCA, Number 2, together account for 50 per cent of total production. GE is third. None of the trio makes picture tubes in the New York Metropolitan Region, though certain smaller firms, including Tung-Sol and Thomas, do make them there.

Receiving and picture tubes do not constitute the entire tube industry. In terms of dollar volume, though not in terms of units, they made up 58 per cent of all electron tubes in 1957.[3] The other 42 per cent consisted of power tubes and tubes for various special purposes. Much of the sub-industry producing these power and special-purpose tubes is located in the New York Metropolitan Region. *Electronic Industries and Tele-Tech* in its June 1957 directory issue lists about 140 manufacturers of tubes of almost 50 different types. About 35 per cent of these manufacturers are in the Region. Many of them are relatively small and cater to a limited market, concentrating on one or two types of tubes which have but few applications. Mass-production techniques are not justified, and the necessary precision hand work makes these tubes very expensive. Some weigh as much as 250 pounds and cost more than $40,000 apiece.

One of the nation's largest producers of power and special-purpose tubes is Amperex Electronic Corporation, now at Hicksville, Long Island, formerly located in the shadow of the Brooklyn Bridge. Sperry Rand has a tube facility at Great Neck, Long Island. One of the leading smaller firms is Anton Electronic Laboratories in Brooklyn which turns out several kinds of special-purpose tubes. The Red Bank Division of Bendix Aviation Corporation in Eatontown, New Jersey, is engaged in making a dozen different kinds of tubes, as is Machlett Laboratories in Springdale, Connecticut.

In the electron-tube industry as a whole, the New York Metropolitan Region grew much more slowly than the nation. Census figures show that between 1947 and 1954, employment in the 17-county Standard Metropolitan Area increased only from 10,000 to 12,000 while national employment was increasing from 28,000 to 71,000. Thus the area's share dropped from 36 to 17 per cent.[4]

Shifts were going on, too, in the manufacture of other types of components, especially capacitors and resistors, though figures are not available. Some of the most important makers of such items had their beginnings in the New York Metropolitan Region. The Dubilier Condenser Corporation, after moving from Manhattan to the Bronx, merged with the Cornell Electric Company to form the present Cornell-Dubilier Corporation, one of the nation's leading producers of capacitors. This firm has been operating since the mid-1930's in South Plainfield, New Jersey—still in the Metropolitan Region—but it has recently built two branch plants in North Carolina. Aerovox was formed in Brooklyn in 1922 by a salesman who sensed an opportunity in the new radio industry; it is now in New Bedford, Massachusetts. Electro Motive started in Manhattan in 1927, but is now in Willimantic, Connecticut, outside the New York Metropolitan Region. Clarostat had its home in Brooklyn until several years ago, but is now in New Hampshire.

These examples could be multiplied. None of the larger manufacturers of capacitors or resistors is now located in the Region except Cornell-Dubilier. Smaller firms in the capacitor field, such as Micamold and Pyramid Electric, still have factories in the New York Region, but have also taken some of their production to southern locations. A Manhattan loft firm, Hammarlund Manufacturing, has transferred most of its production of standard components to North Carolina. Within the Region itself there is a general drift from the central city to the suburbs.

The general story in the location of component-making, then, is that this branch of electronics—or much of it—was originally in the New York Metropolitan Region but today the largest firms have their plants elsewhere. Some important firms, like RCA, have tube facilities in the Region, and many of the smaller manufacturers of power and special-purpose tubes find the area a congenial place to operate. Most of the large firms either moved their component production from the Region or did not choose to locate there in the first place.

Three groups of component firms are left in the New York Metropolitan Region. There are those which cannot muster the resources to leave; there are those which have moved some of their operations to other regions and may continue this movement if conditions so dictate and resources permit; and finally there are those which specialize in turning out custom or nonstandard items.

Firms in this third group are not under the competitive pressures to which the producers of standard components are subject, and furthermore they are considerably dependent on the kinds of advantages which the New York area can offer. Among these advantages, a highly important one is the availability of skilled labor on short notice to fill intermittent needs. Also, firms can obtain specialized materials and parts; they can easily work with machine shops and testing facilities; and they can readily ship to a heavy concentration of electronics markets in the vicinity, including the local producers of a wide range of military and industrial electronic equipment.

The outward movement of other components, it should be noted, has not been principally a drift westward as in the manufacturing of radio and TV sets, but a general dispersal, often into small cities and rural areas, the most notable sections being New England and the South.

Meanwhile the revolution in electronic components—the rapid rise of the solid semiconductor in place of the vacuum tube—is bringing locational implications of its own. But that story belongs mainly to the future and can be better presented in the final chapter.

Spotlight on labor costs

The cost structure of component production is quite different from that of assembling radio and TV sets. Materials make up a smaller proportion of total costs in component plants than they do in set plants. But in the case of labor, it is just the other way around. The cost of labor, relatively small in radio-TV assembly, occupies a position of major importance in component-making. Chart 3 presents the approximate cost structure of six selected components: receiving tubes, picture tubes, power and special-purpose tubes, semiconductors, capacitors, and resistors. The costs are divided into

Chart 3

Estimated Structure of Plant and Distribution Costs for Six Selected Electronic Components, United States, 1957

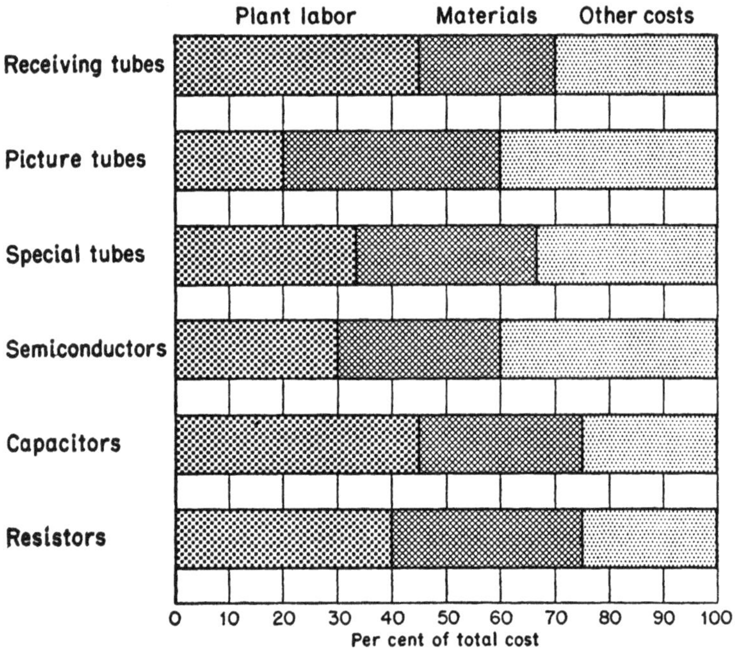

"Special tubes" are power and special-purpose tubes costing over $40.

Source: Interviews with executives of firms manufacturing the components.

three items: labor, materials, and overhead—overhead meaning general factory and distribution expense. Labor runs between 20 and 45 per cent of total costs in contrast to about 10 per cent in set assembly. Materials run between 25 and 40 per cent, in contrast to about 78 per cent in set assembly. Transport costs, both on incoming and outgoing shipments, are unimportant except for certain kinds of tubes.

It is not surprising, therefore, that the most potent force making for shifts in the location of components manufacturing has been the search for cheaper labor. Firms have acquired an increasing need to economize on their labor costs. And this need has put the New York area and other metropolitan areas at a serious disadvantage compared with rural areas.

Why did not the need to minimize labor costs preclude the initial development of manufacturing in the New York Metropolitan Region? The answer is a familiar one. As in radio manufacturing, the technology of component-making had not yet settled down at this early stage and there were some advantages in being close to the center of technical development. This was particularly important because so much of the market for components was concentrated in the vicinity of the New York Metropolitan Region, and the requirements of the users of components played a large role in production. Also in the picture was the special fact that the electric lamp industry provided a pool of workers whose skills were useful in tube-making, as we have seen. The probability is also high that, at the early and uncertain stages in the development of the various components, all of the attractions which the New York Metropolitan Region has for an industry with uncertain prospects and changing technology were operating with their usual force.

With the spread of the market, however, and the stabilization of technology in the large-scale items, the need for a New York metropolitan location weakened. When the technology of tubes and other standard parts had been fairly thoroughly explored (in the 1930's, long before the invention of the transistor), the emphasis shifted to low-cost production. The procedure was to take out an RCA license, get the most modern tube-making machinery, and find a suitable labor supply. Witness the pattern of plant location undertaken by Sylvania, GE, Tung-Sol, Westinghouse, and other firms which "took to the woods" with major portions of their production of standard tubes. All of these companies have sought areas of low labor cost—usually small-city or semirural locations. For example, GE opened a plant at Anniston, Alabama; Sylvania at Burlington, Iowa, and Shawnee, Oklahoma; Westinghouse at Horseheads and Bath, New York.

When television burst on the scene after the war, the companies that had established themselves in receiving tubes became the leaders in picture tubes also, so that picture tubes were never attracted to the New York area in the way that receiving tubes had been. Similarly, it will be recalled, the assembly of television sets was never drawn to the Region to the same extent as radio production in an earlier period. The location of picture-tube plants will be further discussed presently.

The statistical evidence supports the conclusion that component-makers searching for cheap labor could find lower wage scales outside the New York Metropolitan Region than inside it. In the manufacturing of components it is beside the point to compare wages in the New York area with those in the other large urban centers, because in components, much more than in radio and television assembly, the drift has been towards the small town and the rural area. The new locations are mainly in New England and the South, and to a lesser extent in the mining districts of Pennsylvania and West Virginia. The wage differentials in component-making between the New York Metropolitan Region, on the one hand, and New England and the South, on the other, are shown in Table 6, some of the figures for which were obtained by the author from his interviews with manufacturers. The Region's hourly rates run about 15 cents higher than those of New England plants and about 35 cents higher than those in the South. The differences are the most pronounced in the simpler components such as resistors and simple capacitors, and less pronounced in tubes and semiconductors.

Table 6 Lowest and Highest Wage Rates and Estimated Range of Average Hourly Earnings in Components Manufacturing for New York Metropolitan Region, New England, and the South, 1957

	Lowest wage rate	Highest wage rate	Range of average hourly earnings
New York Metropolitan Region . .	$1.10	$1.80	$1.40–1.60
New England	1.00	1.60	1.25–1.45
South	.80 [a]	1.30	1.05–1.25

[a] Learners' permits. Some employers have received exemptions from the federal minimum wage for learners.

Source: *Hearings,* "Learners in the Small Electrical Products Industry," Wage and Hour and Public Contracts Divisions, U.S. Department of Labor, 1956, passim, and figures obtained by the author in interviews.

Nor is there much question about the reasons for the lower wages in those other areas. The supply of unskilled labor has been much more plentiful there, because the birth rate has been higher, agriculture and other extractive industries such as coal mining have declined as sources of employment, and, finally, female labor in rural areas and small towns has found fewer alternatives to factory employment than female labor in large metropolitan centers. Negro and Puerto Rican migration to New York has not been sufficient to match the low-wage labor supply of the agricultural areas.

The discrepancy in wage levels between the New York Metropolitan Region and areas of labor surplus has not provided a strong enough force to draw away many industries in which labor costs are only a modest proportion of total costs, or in which the need is for skilled labor. But in components manufacturing, as we saw, the labor content is high and the great majority of the workers are female. The attraction of low-wage areas is therefore quite strong in this industry.

Wages, of course, are not the only elements of labor cost. Others are fringe benefits (paid vacations, pensions, insurance, and so on) and productivity levels. But the costs of fringe benefits, like wage costs, are generally higher in the New York Metropolitan Region; and the lower levels in rural and small-town areas thus tend to strengthen their attraction. Even more important is the fact that fringe-benefit costs build up as time goes on because of seniority privileges; when a firm moves, it acquires a new crop of employees without seniority, and the cost of fringe benefits decreases for that reason even if for no other.

The question of productivity is a little less clear. Neither lower wages nor lower fringe benefits would be compelling reasons for manufacturers to seek out areas of labor surplus if those gains were dissipated in lower productivity. Despite the unskilled nature of the work, a certain amount of training is necessary, and we may reasonably ask whether the cost of training makes a serious dent in the savings resulting from lower wage and benefit costs. The evidence on this point is somewhat mixed. What seems to be the case is that even experienced factory workers need some training, but that some employers have not had to bear the entire cost because they have obtained temporary exemptions from the federal minimum-wage law ($1 an hour) which permitted them to pay as little as 80 cents an hour for a period of 12 weeks.[5]

Aside from the training period, it would seem that productivity in the low-wage areas is at least as high as, or higher than, that of the New York Metropolitan Region. Over the years, in the areas where many of the large component-makers were originally located, continual re-rating of jobs in accordance with increased mechanization led inevitably to labor strife. Thus, by moving to new areas where the labor force is unfamiliar with the traditional standards, manufacturers can often attain higher levels of output per man-hour along with lower wage scales.

In addition, the manufacturer is likely to be the only employer of factory labor in the area from which he draws his help, and if so, he gets the cream of the supply. One employer who opened a plant in the South reported that 3,700 people applied for 250 jobs. On the other hand, in metropolitan centers, other manufacturers are constantly outbidding the component-makers for the best of the work force.

OTHER COST FACTORS

Though the cost of labor has been the strongest locational factor in the components field, other factors have shown great strength for some kinds of components.

It will be recalled that in discussing the assembly of sets we emphasized the importance of transportation costs. These have also been a critical factor in the location of plants making picture tubes. The glass "blank" which goes into the making of a picture tube is a fragile and costly item to ship; and so is the finished tube itself. Manufacturers of tubes, therefore, have an incentive either to locate close to the sources of the blanks or close to the manufacturers who buy the tubes, thus eliminating one of the perilous and costly freight journeys.

For example, Sylvania and Westinghouse have plants in upstate New York which draw their blanks from the nearby glass works at Corning. CBS-Hytron built a picture-tube plant in Kalamazoo, near a source of blanks in Albion, Michigan.[6] On the other hand, RCA makes its picture tubes at Marion, Indiana, close to its TV assembly plants; and Sylvania built a picture-tube plant in California, remote from plants producing glass blanks, in order to be near the California set-makers and the big replacement market in that state. (About 30 per cent of all picture tubes produced in the United States are sold as replacements.)[7] Some picture-tube plants, in Ohio, for example, manage to be fairly close to both glass works and set-manufacturers.

All such locational decisions are increasingly colored by the role of the reworked picture tube. When tubes fail, the glass is often used again to build new tubes; thus the households of every metropolitan center are both a source of glass and a market for the tubes produced with the glass. Therefore centers of heavy population exert another locational pull upon the producers of tubes, a pull which probably explains in part Sylvania's location of a branch plant on the West Coast and the location of many small tube-working plants in metropolitan areas.

Collectively, all these transportation forces complicate the tube-maker's locational calculus; the cost of labor is far from being his sole criterion, but must be balanced against compelling transportation needs.

Aside from the glass required for tubes, the materials used in the manufacture of components are widely available at uniform prices and their cost does not prevent a producer from going where labor is cheapest. Nevertheless, if producers respond to opportunities to cut labor costs in small towns and rural areas, they do sacrifice some advantages, such as that of proximity to one's own suppliers, permitting a more flexible inventory policy. Another is proximity to services. The ability to deliver more quickly than a competitor may depend on the swiftness with which patternmaking or tool-and-die work can be accomplished in the shops of subcontractors. In a rural area, even if such outside services are available they may not be readily expansible, and the component firm may have to wait its turn and perhaps lose the order to a manufacturer who can "turn on a dime."[8]

Materials and services are not the critical locational factors in the making of standardized components. In such a competitive business, however, a slight advantage in any element of cost can assume unexpected importance; and the desire for flexibility must be considered a locational force somewhat retarding the movement toward the available pools of cheap labor.

Indeed there is one kind of component-maker for whom the flexibility factor is critical. This is the manufacturer of unstandardized components, like special-purpose tubes. Because he must have specialized materials and services, and because his product is more subject to rapid change, the attraction of low-wage areas is not compelling. The essential raw materials and purchased parts used in the manufacture of the common components are widely available, but the plant turning out unstandardized or custom-built components must have materials that meet exceptionally rigid specifications in terms of chemical composition or physical properties. And such materials are readily available only where demand justifies providing them on other than a special-order basis. Such demand is usually in urban centers. For example, most of the makers of special-purpose tubes housed in the New York area require a whole range of metal parts which only skilled metal workers can turn out, and the envelopes for such tubes require the services of expert glass blowers. Accordingly, one could hardly expect a firm making special-purpose tubes to locate in a North Carolina hamlet.

Work in metallurgy pertaining to electronics is being done at centers of research such as New York University, Massachusetts Institute of Technology, California Institute of Technology, and Carnegie Institute of Technology. Much of it is being subsidized by the federal government. Firms working on specialized applications are in the metropolitan areas surrounding such centers. Since Edison's heyday one of the noticeable attractions that the New York Metropolitan Region has exerted on electrical and electronic production has been the magic of research.

The Region's attraction for producers of unstandardized components also has to do with the kind of personnel they require. The engineer and the highly skilled laborer have a more essential role in these firms than they do in the making of standard components. The firm's demand for such personnel is likely to fluctuate, both in terms of the variety and quantity needed. Whereas the small town and rural area can furnish unskilled labor, a metropolitan area has the edge in catering to the fluctuating demands for professional and skilled labor. To some extent, the old mill towns of New England with their heritage of skills have been able to accommodate more complex operations than certain other areas of surplus labor could accommodate. But for products which require sizeable expenditures on research and development and much skilled labor in their manufacture, facilities have remained in use—or have been established—within the large metropolitan areas including the New York Metropolitan Region. In this propensity for metropolitan locations the production of unstandardized components is very like the third major branch of the electronics industry, military and industrial electronics.

4

Military and Industrial Electronics

A missile climbs into the sky at Cape Canaveral. Guidance controls take it toward its destination, while telemetering equipment reports its progress. Electronics must "see, hear, feel, and measure every manner of physical phenomenon in its path; analyze the data perceived and adjust accordingly; transmit the information back to its launching base; and receive further instructions. . . ."[1] As the missile speeds through space, American military aircraft are aloft in many parts of the world, each one carrying a ton or more of electronic apparatus. Electronics is also at work in land installations and on submarines. Radar extends military vision far beyond the abilities of the human eye and under conditions of darkness, clouds, or fog. Computers have almost removed the human element from the sighting and firing of large weapons. All this is a considerable change from the day in 1899 when the Army Signal Corps succeeded in transmitting a wireless signal between Fire Island and Fire Island Lightship, off the Long Island coast.[2] And it is a change of economic as well as military significance. Here we are concerned with the economic aspects, and particularly the question of where the equipment is manufactured, and why.

We lump together as "military and industrial electronics" the residual part of the electronics industry, that is, everything except consumer electronics and components. The term "residual," however, belies the importance of this third category, for it has expanded immensely since World War II, and by 1957 it accounted for more than two-thirds of the dollar value of all electronic products. Military items alone, including missile-guidance systems and radar, accounted for more than half of the entire electronics industry, by value of product.[3]

We use the broad term "industrial electronics" to cover the many kinds of commercial and industrial electronic devices used by private companies, institutions, and governments in factory production, office work, public services, and research. Here are a few illustrations adapted from a recent publication of the State of New York. Two-way radio systems are multiplying so rapidly that the total number of federally licensed transmitters is close to a million; such systems are found not only in taxicabs and police cars but also on construction jobs, in mines, farms, stores, factories, and warehouses. Facsimile, often called "fax," is used to transmit newspapers, drawings, charts, and financial statements, signatures for verification, weather maps for airlines, and Pullman reservations. Television is used industrially for communications, training, and "monitoring"—that is, keeping a watchful eye on an atomic reaction, a water gauge, a plant gate, the floor of a department store, a procession of freight cars, or the level in a two-story tank of paper pulp. "Electronic eyes" open doors, warn when smokestacks emit too much smoke, count shoppers, and synchronize conveyors. Electronic testing and inspection equipment of some kind is in daily use in thousands of factories and practically every industrial and scientific laboratory.[4]

Some products of course have both military and private uses—for example, aircraft systems and computers. A computer weighing 275 tons and containing 58,500 vacuum tubes is at the heart of the nation's air defense system. And in the nation's offices and research laboratories, substitution of electronic computers for clerical labor is proceeding at such a pace that a four-fold increase in sales of such machines between 1957 and 1965 could be conservatively predicted.

With every passing year the frontiers of electronics are extended to embrace additional fields both in industry and in national defense. The subject of the present chapter is therefore of utmost importance to our investigation of the forces affecting the location of electronics production. A projection of the future role of the New York Metropolitan Region will depend to an important degree on one's guess as to whether the Region will hold and attract firms making products in this category.

Location of the plants

It is in military and industrial electronics that the New York Metropolitan Region has so far displayed the greatest staying power. The Region has consistently been the nation's leading center by a wide margin. Comprehensive statistical measures for this rapidly growing and diversified sector of electronics are even scarcer than they are for set assembly and component-making. But we can draw some indications from the limited information which is available.

A special report of the federal Bureau of Labor Statistics for 1951 showed that the Standard Metropolitan Area led the nation in that year with 17.6 per cent of the national employment in military and industrial electronics. Philadelphia accounted for 10.6 per cent and Los Angeles 9.8 per cent.[5] So much has happened since 1951, however, that we could hardly afford to rely on these figures to ascertain the current position of the New York Metropolitan Region.

In 1954, when the last Census of Manufactures was taken, the Region, as represented by the slightly smaller Standard Metropolitan Area, accounted for 38.6 per cent of employment in the industry called "Scientific Instruments." Much of this industry consists of military and industrial electronics production, as illustrated by the fact that in 1954 about three-fifths of the value of shipments of "Scientific Instruments" was in aircraft flight instruments and automatic pilots, a large portion of this gear being electronic.[6] The same census showed an increase in the Region's share of the industry called "Radios and Related Products," an industry which embraces both consumer and military-industrial electronics. We can be reasonably certain that the observed improvement of the Region's position is a reflection of its strength in the military and industrial category.

In the absence of precise statistical measures, even such a crude means as the scanning of industrial directories can be helpful in assessing the Region's role in military and industrial electronics. The 1957 directory issue of *Electronic Industries and Tele-Tech,* for instance, lists the firms in the nation with capabilities in specified fields. The "capabilities" concept is highly subjective, of course, but the lists show that for most major military and industrial products, something like one-third or one-fourth of the nation's firms with "capabilities" for manufacture are located in the New York Metropolitan Region. Proportions of this sort appear in such products as special-purpose amplifiers, computers, microwave components, radar devices, auxiliary electronic equipment for aircraft, and many others.

In general, this branch of the industry has favored locations in large metropolitan areas. Eight areas accounted for 73 per cent of the nation's employment in 1951.[7] All of these areas except Syracuse had populations of over 1,000,000; and among them were the four largest metropolitan areas in the country. This strong orientation to large metropolitan areas is by no means typical of most manufacturing industries. Moreover, this orientation does not take the form of a multitude of small firms, concentrated in the central city, as in the case of apparel. If we again take the Census industry which we have viewed as very roughly approximating military and industrial electronics, namely "Scientific Instruments," we find that only a small fraction of its employment in the Region is in New York City and that the average plant in the Region is much bigger than the average plant in the nation.

In military and industrial electronics, the Region is noted for large firms and large establishments, examples being International Business Machines at Poughkeepsie (and at Kingston, which is nearby but just outside the Region); General Precision Laboratory, Inc., at Pleasantville in Westchester County; the Norden-Ketay

Division of United Aircraft Corporation in Fairfield County, Connecticut; two subsidiaries of International Telephone & Telegraph Company in New Jersey; and Fairchild Camera and Instrument Corporation and Airborne Instruments Laboratory in Nassau County, Long Island.

Some of the firms now operating in the outlying counties of the Region were formerly quartered in New York City. The Kearfott Company, which produces a wide variety of precision components for military equipment in eight plants in the Paterson-Little Falls-Clifton area of New Jersey, began by making ship windows in Manhattan in 1917. The large plant of American Bosch-Arma at Roosevelt Field on Long Island is the outgrowth of what was once a Brooklyn manufacturer of searchlights for the Navy.

New York City still has some large plants in the military-industrial category.[8]

Radar equipment is made in a plant of the Otis Elevator Company in Brooklyn, and by the Ford Instrument Company Division of Sperry Rand, in Queens. There are other important installations in Queens. A plant of the Skiatron Electronics and Television Corporation in Manhattan has been producing transmission systems to be used in conjunction with radar. Loral Electronics Corporation in the Bronx possesses a wide range of capabilities in military electronics, having grown from a phonograph producer.

Of course the Region also contains many small companies which are doing work in military or industrial electronics. Many of these employ only a handful of people. Others are somewhat bigger, such as Hammarlund Manufacturing Company and W. L. Maxson Corporation, both of which are engaged in military work in Manhattan. The limited evidence at our disposal suggests that the majority of these small plants—or at least many of them—are located in New York City.

In short, both large and small plants have found the Region a congenial place. Presumably the same is true of other metropolitan areas which are strong in military and industrial electronics. What accounts for the dominance of the large metropolitan areas? And why does the New York Metropolitan Region rank first by a substantial margin?

In accounting for the rapid spread of defense electronics on Long Island, and in the whole Region for that matter, one is tempted to jump to the conclusion that this growth is largely attributable to the demands of the nearby aircraft and missile manufacturers, such as Republic and Grumman. It is certainly true that these aircraft manufacturers do draw electronic products from electronics firms in the Region. But these electronics firms also supply manufacturers all over the country. This is clearly suggested by the results of a transportation survey summarized in Table 7. In fact, the Region's military and industrial electronics plants ship a larger proportion of their products to distant markets than do other "national-market" industries in the Region. The firms in the sample ship 21 per cent of their product, by weight, to the Pacific Coast and 22 per cent to the South. Thus, the Region's electronics producers not only cater to the needs of nearby aircraft manufacturers but also participate to a large extent in meeting the needs of aircraft and missile producers on the Pacific Coast and in the South. To explain the strength of

Table 7 Average Percentages by Weight of Outgoing Freight for Selected Manufacturers of Military and Industrial Electronics Apparatus in the New York Metropolitan Region, by Destination, 1957

New York Metropolitan Region	8
Middle Atlantic outside NYMR	10
New England	10
East North Central	16
South	22
Mountain and Plains states	12
Pacific Coast	21

Source: Derived from a study of transportation in the New York Metropolitan Region, to be reported by Benjamin Chinitz in a forthcoming volume in this series.

the Region in the military and industrial fields we must look at other factors.

The need for flexibility

Though the demand for military and industrial electronics has grown large, it is extremely unpredictable. As long as the cold war persists there will be a high rate of defense spending, but a Korea or a sputnik can lead to quick acceleration, while a squeeze by the national-debt ceiling can work in the opposite direction, as many firms found to their dismay during 1957. Even in the industrial field there is considerable instability because of the vagaries of the business cycle. The purchase of electronic equipment, like all capital spending, is subject to wide fluctuations in response to swings in business expectations.

When viewed from behind the desk of an electronics manufacturer the instability of demand is magnified by the rapid rate of obsolescence and the highly unstandardized nature of the product. Even if the Defense Department were to make an irrevocable commitment to spend billions annually, the basket of goods purchased would consist of a tremendous number of different items, very few of which would be ordered in large volume. Furthermore the contents of the basket would be revised frequently. As potential enemies discover ways of neutralizing our offensive power we invent new offensive weapons; as they invent new offensive weapons we have to discover ways of neutralizing those.

Our constantly changing armory precludes the stabilization of commercial relationships between the government and its suppliers such as prevails between private buyers and sellers. The military services encourage invention and adaptation through their research-and-development contracts, but at the point of large-scale procurement they must choose the product which will do the job best.

In such a market situation it is necessary for the firm to maintain a high degree of flexibility in its operations. This means being able to respond quickly to the fluctuations of the market; to expand, contract, and diversify at maximum speed with minimum financial penalties. Such flexibility is achieved by keeping fixed costs to a minimum and being poised to enlarge capabilities with the aid of materials and services purchased in varying increments for varying time periods.

One might be led to suppose that the uncertainties associated with this branch of electronics would preclude the establishment of large plants. But a firm usually must be large to compete for government contracts which involve the design and production of expensive and complicated equipment. In other words, though the risks inherent in this business inhibit the attainment of self-sufficiency, the technology precludes an industrial organization in which all the firms are small. What we have instead is a group of large firms with substantial resources; each of them is capable of assuming responsibility for a big contract of the type found in the procurement of weapons systems; but the large firm relies, in turn, very heavily on many specialized small plants to fill the gaps in its own capabilities.

The industrial environment of a large metropolitan area has much to offer to firms, both large and small, which need to operate in the kind of framework we have been describing.

The advantage of a metropolitan location for both "primes" and "subs"—prime contractors and subcontractors—lies not only in the need that each firm has for suppliers, but also in the need to be in close touch with them. This need grows out of technical difficulties in the product itself. The rapid advances in research and development, keyed to the changing requirements of both military and industrial electronics, make geographical propinquity of primes and the several levels of subs desirable for two reasons.

First, it is usually impossible to meet specifications merely by hooking up sub-assemblies and components that are already commercially available. Constant liaison must be maintained between prime and sub to develop new types to meet the demands of space, weight, environment, and reliability. Design and testing of these items must be carefully meshed with the progress of the complete system. It is true that such meshing can be done when the prime contractor and his subcontractors are geographically separated, and often is. But the tasks are much simplified when face-to-face contact is possible on a daily basis. Much of the work in the New York Metropolitan Region is of this nature. Even large manufacturers, therefore, may be drawn to the concentration of firms capable of doing such specialized work, whose talents lie in different fields.

Secondly, propinquity is desirable in yet another way. The subs find it advantageous to be near one another. It is basically a matter of information, both about technological changes and about the opportunities to cater to them. In addition, the subs work for one another, in what amounts to a trickle-down process.

The dependence of the small firm on the outside services available in metropolitan areas is being explored in a number of volumes in this series. The category of military and industrial electronics gives us an excellent example of this dependence—in this case dependence on testing laboratories. Such laboratories often serve many industries and tend to locate where industries are clustered. An executive of one of the largest testing laboratories in the New York Metropolitan Region, located in Manhattan, reports 3,500 customers, most of which are small and medium-sized firms and many of which are engaged in some aspect of electronics production.[9]

As we have seen, the volume of business a firm can expect to do in this category of electronics is not easily predictable. "Feast or famine" is the phrase often used by executives. This uncertainty requires much flexibility in plant facilities. The Manhattan loft districts offer such flexibility. But even firms located on Long Island, in Westchester and Fairfield Counties, and in New Jersey suburbs often occupy rented quarters. The managements wish to be in a position to make a profit whether the volume of business is 15 or 65 million dollars per year. In their view it is impossible to schedule far in advance the type of work they do, as it is so closely connected with contract awards from governmental agencies. Not only this, but in military electronics and in certain kinds of industrial electronics, design and development must precede manufacture;

and the time taken by design and development is usually an unknown quantity. A lot of physical facilities on which commitments had been made could therefore easily amount to a serious financial burden.

ENGINEERS AND SKILLED WORKERS

Throughout this study we have emphasized the importance of the cost structure of electronics firms as a guide to their location preferences. In military and industrial electronics the outstanding feature is the heavy use of engineers and highly skilled workers. For example, according to the latest available figures, engineers were less than 2 per cent of the labor force in radio-TV and component plants but were 9 per cent in military and industrial electronics plants.[10] The need for trained personnel has had strong influence on the location of these establishments.

Engineers. The concentration of military and industrial electronics plants in the New York Metropolitan Region benefits both the engineer and the employing firm. The individual, who is a specialist in one of the complex fields of electronics, has many potential job opportunities; the firm, which employs specialists in various fields, has many potential candidates. Furthermore, once placed, the individual has the freedom of mind which comes from the knowledge that he can move to another job in the area with but little difficulty. On the other hand when an engineer is employed by a firm in a so-called "country" location—"country" in this case meaning everywhere except the largest metropolitan areas—a change of jobs will usually entail the expense and inconvenience of moving his family. Even though in the metropolitan environment his job may be "insecure," in that the firm has made no commitments to him, he may well accept this disadvantage to avoid the problems which could result in the country location. Engineers have even been known to turn down higher salaries in order to "stay in the market."

The firm, too, has its choices to make. Being in the New York

Metropolitan Region gives it flexibility in that it may hire personnel to whom no long-term commitments have to be made. But it must also live with the knowledge that the existence of nearby employment opportunities, which its employees are sure to learn about through newspaper advertisements and trade gossip, can result in an unstable work force, bringing with it the costs of labor turnover. Nevertheless, a flexible personnel policy is possible in such a market and is desirable to a firm whose volume of business fluctuates both in dollar volume and range of products. Furthermore, the firm in the Region has greater freedom to initiate frequent changes not merely in the *size* of the staff but also in its composition. In a country location the firm is inhibited by recruiting expenses and by commitments made to the staff; it feels a strong need to maintain a reputation as a benevolent employer. In a sense, then, the firm becomes dependent on its professional talent.

Personal and social considerations are often as important as the economic and financial considerations just reviewed. For example, a vice-president of a company in the New York area said that in establishing a branch operation in a small southern town, he had great difficulty in getting persons with Jewish and Roman Catholic backgrounds to move away from New York. The heterogeneity of the population in the New York Metropolitan Region makes it possible for members of almost any religious or ethnic group to find a congenial environment there.

The Region also offers environmental congeniality in the sense of intellectual stimulation, whether from the standpoint of education *per se*, or the availability of one's professional cohorts with whom to discuss matters of mutual interest, or the enjoyment of cultural activities as a whole. The younger engineers have opportunities to continue their education with hopes of qualifying for a master's or doctor's degree. New York educational institutions, such as New York University, Columbia University, and Brooklyn Polytechnic Institute, afford these opportunities. Other metropolitan areas also offer good colleges and universities, but most country locations do not. Art galleries, theaters, the opera, concerts, and lectures contribute to the attraction that New York possesses for professional groups. If one is accustomed to these things, a firm in a country location may appear to the prospective employee to be in an intellectual desert.

Age differences among engineers help to determine their locational preferences. The young man feels a greater need to be near educational institutions. The older man, already possessing the degrees he feels he will need, may prefer the less hectic pace and the amenities of the so-called country locations—but even this attraction is being counteracted by the decentralization of industry within the metropolitan areas, for he can enjoy the amenities and still live within easy distance of his work.

In terms of where engineers are "made," the New York area unquestionably leads the nation. As shown in Table 8, New York State had 33 per cent of the nation's graduate students in electrical engineering in 1954, and New Jersey had another 10 per cent. Thus the two states together had 43 per cent, and most of these were trained in the New York Metropolitan Region.

Table 8 Resident Graduate Students in Electrical Engineering in the Seven Leading States, April 1954

	Number of students	Per cent of United States total
United States total	4,959	100
1. New York State	1,623	33
2. Massachusetts	580	12
3. New Jersey	520	10
4. California	403	8
5. Pennsylvania	358	7
6. Maryland	251	5
7. Illinois	222	5
All other states	1,002	20

Source: *Graduate Student Enrollment and Support in American Universities and Colleges,* 1954 (Washington: National Science Foundation, 1957), pp. 238–239.

The importance of the Region as an engineering center can be illustrated by the case of Sylvania Electric Products. This firm has its basic research and experimental laboratories at a number of locations in the New York City borough of Queens, even though it carries on its principal manufacturing operations in small towns and cities far from the Region, and even though it produces military and industrial electronic apparatus at several places in the industrial complex surrounding Boston.

Many executives of firms in the New York Metropolitan Region, when interviewed by the author, said these firms would not have facilities in the area if it were not for their dependence on engineering talent which would be difficult to obtain and hold in other locations. Many said their production costs in the Region were above those of other areas, and that a location in the Region did not put them particularly near their markets or their supplies of materials. With a sustained easing of the general shortage of engineers, such firms might be expected to think seriously about moving some or all of their operations.

Skilled labor. In military and industrial electronics, skilled workers are required not only in relatively large numbers but also in considerable variety. In the design and construction of radar equipment, for example, a good deal of knowledge about mechanics is needed in the original layout of the project and then in construction. In addition, a well-staffed machine shop is customarily a necessity. Many technicians whose talents fall between engineering and skilled bench labor are used at all levels to assist engineers, to perform tests on the "line," and to run extensive tests on the completed products. The demands for reliability of products are most stringent, so that thorough inspection procedures are essential. Technicians are also needed to keep the testing equipment itself in operation.

No firm can expect to find a pool of such varied talents in places where these talents have not been employed over the years. Flight to low-cost labor areas, therefore, is often impossible for firms in military and industrial electronics, because the kind of labor they require is not the kind which is available there in large quantities. Over the long run, firms can train their own help for specific tasks of a technical nature, but this is expensive. Certificates of necessity—which permit liberal depreciation allowances on taxes—have been granted for the construction of facilities in such areas, but some firms which have been lured by this subsidy have experienced serious initial labor shortages.

The Region's decline as a center for the assembly of radio and TV sets and for the making of components has in a sense contributed to its growth in military and industrial electronics. Many firms in the Region, faced with formidable competition from outside plants in the more traditional fields, have been more than willing to explore the opportunities on the rapidly expanding scientific frontier. They were able to cash in on the pool of talents and the experiences of an existing organization. Without exception, the larger, well-known firms in consumer electronics in the New York area, such as Emerson, DuMont, and Olympic, have in recent years gone into military and industrial electronics.

THE RISE OF OTHER AREAS

We have seen that the military and industrial category of the electronics industry has grown so fast that in dollar value of product it has surpassed consumer electronics and components put together. This tremendous growth has taken place in various parts of the country. The limited evidence at our disposal suggests that the New York Metropolitan Region has maintained its leading position in military and industrial electronics because of the uncertainty of the market, the unstandardized nature of the product, and the heavy reliance on engineering talent and skilled labor. These characteristics have favored large metropolitan areas, and particularly the New York Metropolitan Region where military and industrial electronics plants are both large and small, with prime contractors and subcontractors living in a complex state of interdependence and benefiting by being near one another.

Yet the strengths of the Region seem to be only strengths in transition. In the first place, even in military and industrial electronics, some products have become relatively standardized—for example, electronic eyes, electronic counters, and certain measuring instruments. Indeed, products are always changing in this respect and, whenever one of them reaches a point at which production can be accomplished largely by unskilled labor doing the same tasks over and over, this production tends to break off from the metropolitan setting and go where labor costs are lower. In the second place the very advantages which the Region offers for unstandardized products—the pool of engineers, the abundance of services and suppliers—are gradually being duplicated elsewhere.

Accordingly the industrial structure of this branch of the industry—large and small plants depending on one another—can be seen in various stages of development in other areas of the nation. The West and South are particularly notable. The tremendous expansion in the West is linked directly to the aircraft industry, though some eastern manufacturers have established branch plants there to supply other electronics users. The eleven states in the Pacific and Mountain regions claimed 23 per cent of total electronics sales and 18 per cent of the industry's employment in 1957.[11] The recent rising importance of various sections of Florida and of the region around Denver can be linked to military work. If the Martin Company had not set up its important missile work in the Denver area, other firms would have been unlikely to locate electronics plants there. In the industrial field, work at General Mills and Remington Rand in the Minneapolis area has created an atmosphere conducive to the further establishment of electronics facilities.

Such centers of electronics development generate a demand for specialized services which the pioneer facilities must at first provide internally. As the demand grows, companies find it profitable to specialize in providing these services. For example, as electronics firms multiply in the San Francisco area, another firm builds a plant to design and fabricate sheet metal parts for the electronics industry.[12]

The boom in Florida is one of the most remarkable of all. The state has created a favorable business climate, and the natural climate makes its own contribution. Industry has succeeded in drawing substantial amounts of technical and skilled labor to the area, though there are also many cases in which scientists have refused to leave the New York Region even for Florida's sunshine. Such firms as Sperry Rand, General Electric, Minneapolis-Honeywell, and others are in the Florida picture. Some fifteen or more missile contractors maintain research facilities near the rocket base at Cape Canaveral. Here again a few large plants serve as a nucleus for growth and are followed by smaller firms which cater to their specific needs or begin a subcontracting relationship with their larger neighbors.

Florida, however, cannot approach the magnitude of electronics in California, where there are the attractions of educational opportunities, research activities, and natural climate as well as an unusual *esprit* in the business community, evidenced by the West Coast Electronics Manufacturers Association. Southern California's growth in electronics began some time ago, because the area has long been the principal center of the aircraft industry. As electronics has become more and more important in aircraft, a concentration of military and industrial electronics firms has developed. It has been estimated that at least 75 per cent of electronics in Southern California is dependent on defense, and that in nonentertainment electronics this area is second only to the New York area.[13] The rapid expansion of activity in the San Francisco area has been more recent. The emphasis on research there is probably even heavier than in Southern California. Research organizations in the area include Stanford Research Institute, Broadview Research Corporation, Cascade Research Corporation, and the Fisher Research Laboratory.[14] In addition there are company research facilities of many of the "who's who" of the electronics world, particularly in instrumentation, communication, and tube work.

A look around the country tells us, then, that many of those qualities possessed by the New York Metropolitan Region are appearing in other areas. Military work in particular has exhibited a tendency to grow fast in the neighborhoods of its end use—near the aircraft plants, the missile manufacturers, and the testing sites. Nevertheless the military and industrial category is growing so swiftly that if the Region is unable to retain its present share, it still may experience a large absolute increase in jobs.

5

The Shape of Things

Three broad tendencies have been identified in the last three chapters. Production of standard consumer electronics products, which are principally radio and television sets, has been gravitating away from the eastern metropolitan centers to midwestern locations—generally to other metropolitan centers but also to more isolated places. The reason for their leaving New York turns out to be labor cost in part, and the factor pulling them to the Midwest turns out to be the cost of transporting the product. The manufacture of standard components has also left New York to escape high wages, settling in lower-wage areas like New England and the South. In both these categories of the industry we detected strong inertial forces tending to preserve the relative position of the New York Metropolitan Region for particular products, such as nonpackaged hi-fi equipment and special-purpose tubes. And these inertial forces —including a variable demand for engineering talent and skilled labor and the interdependence of establishments—were deemed critical in maintaining the dominance of the Region in the third category of the industry, namely, military and industrial electronics.

Running fast to stand still

What has been the net effect of these tendencies upon the position of the Region in the electronics industry in recent years? To help answer this question it will be necessary once again to approximate the electronics industry by adding up the employment in "Radios and Related Products," "Electronic Tubes," and "Scientific Instruments," the three Census industries which come closest to covering the range of products considered in this study. In comparing New York with the nation we are obliged to use figures which apply only to the 17-county Standard Metropolitan Area. Table 9 shows

Table 9 Distribution of United States Employment in the Three Major Census Industries Producing Electronic Items, 1947 and 1954

		Percentage shares of U.S. employment (U.S. = 100)				
	Number of employees, U.S. total	New York metropolitan area[a]	Rest of Middle Atlantic	New England	North Central regions	South and West[b]
All three industries						
1947	224,717	21	23	10	38	8
1954	410,263	18	22	11	33	16
Radios and related products						
1947	178,595	16	23	10	44	7
1954	293,998	15	22	9	37	17
Electronic tubes						
1947	27,703	35	31	15	10	9
1954	70,950	16	32	21	17	14
Scientific instruments						
1947	18,419	45	21	3	16	15
1954	45,315	39	5	8	34	14

Note: Not all products of these industries are electronic; and some electronic products are found in other industries.

[a] Standard Metropolitan Area as used by U.S. Census Bureau.

[b] This column covers the following Census regions: South Atlantic, East South Central, West South Central, Mountain, and Pacific.

Source: U.S. *1954 Census of Manufactures,* including unpublished special tabulations.

that between 1947 and 1954 the area's share of national employment declined from 21 per cent to 18 per cent. Meanwhile, major growth occurred in the South and West. The absolute increase in national employment in the industry was so rapid, however, that the New York area, despite the decline of its *share,* increased its employment by more than 50 per cent—from 47,000 to 73,000.[1]

Although the individual Census industries do not correspond to the categories of the electronics industry which we have been employing, it is nevertheless possible to compare our earlier observations with the Census figures shown in Table 9. The Census industry "Electronic Tubes" corresponds closely to a major part of our "components" category, and the decline of the New York area's share, it will be noted, is quite severe. The very modest decline in the case of "Radios and Related Products" undoubtedly reflects the balance of two opposing trends: on the one hand, the failure of the New York area to grow as rapidly as the nation in the manufacture of radio and TV sets and components like resistors and capacitors; and, on the other, the area's retention of its strong relative position in military and industrial electronics. Since factory sales of the latter branch of the industry grew much more rapidly in that period than consumer electronics, the net effect has been to leave the area's share almost unchanged.

What has happened to the distribution of electronics employment within the New York Metropolitan Region? Here the data are a little more recent and more complete.* The figures are shown in Table 10. New York City's share of the Region's employment declined sharply between 1947 and 1956, from about 36 to about 24 per cent. Nassau and Suffolk Counties and the New Jersey counties of the Region have made the greatest advances. Again, each part of the Region, including New York City, registered impressive absolute growth, reflecting the rapid expansion of the industry.

* The 1956 data shown in Table 10, based on state employment figures, appear to suggest a much more rapid postwar growth for the Region than the 1954 data shown in Table 9, based on federal Census of Manufactures data. Part of the difference is explained by the different dates, since considerable added growth occurred between 1954 and 1956. Another part of the difference is due to the fact that the 1956 data cover the 22-county New York Metropolitan Region, whereas the 1954 data cover the 17-county Standard Metropolitan Area (SMA). But some of the difference is also due to vagaries of different reporting systems. Since the federal and state figures are sometimes incompatible, even for the same year, it has been our practice to use the federal figures, which apply to the SMA, whenever we are comparing with national figures. It should be noted that the New York Metropolitan Region (but not the SMA) encompasses fast-growing Dutchess County which contains a major plant of International Business Machines.

Table 10 Distribution of Employment Within the New York Metropolitan Region in the Three Major Census Industries Producing Electronic Items, 1947 and 1956

	Number of employees, NYMR total	Percentage shares of NYMR employment (NYMR = 100): New York City	Nassau, Suffolk	Other New York and Connecticut counties	New Jersey counties
All three industries					
1947	44,251	35.8	14.0	10.9	39.3
1956	98,290	23.5	20.8	4.2 [a]	51.5
Radios and related products					
1947	26,221	53.8	1.7	16.5	28.0
1956	54,410	30.5	5.6	4.5	59.4
Electronic tubes					
1947	2,665	7.4	....	0.7	91.9
1956	6,347	14.0	4.5	2.3	79.2
Scientific instruments					
1947	15,365	10.0	37.4	3.2	49.4
1956	37,533	15.0	45.6	4.1	35.3

[a] Decline here is entirely in Fairfield County.

Source: These are our estimates based on unpublished data collected by state departments of labor under unemployment insurance programs for 3rd quarter 1947 and 1956, except for Fairfield County, Connecticut. The Fairfield employment figures are based on *County Business Patterns, First Quarter 1947,* and on unemployment insurance data for 1st quarter 1956.

Critical factors of location

We intimated at the outset that the study of the electronics industry might prove helpful in suggesting critical forces affecting the growth of industry in the New York Metropolitan Region. Of greatest significance is the insight which this case study offers into the sensitivity of location to changes in demand and technology.

Locational shifts are commonly analyzed in terms of changes in the competitive strength of various areas. We too have dealt with such changes, discussing the impact on the Region of developments in labor and transport. But our primary emphasis has been on the changing nature of the industry itself.

Location in the New York Metropolitan Region has been viewed here as a passing phase in the development of an industry. Inherent in the introduction of new products are uncertainties both of production techniques and of prospective demand. Faced with these uncertainties, firms tend to limit their capital outlays and to direct their sales efforts toward the large concentrated markets. Uncertainty in techniques tends to attract them toward metropolitan areas where they can best operate on a "hand-to-mouth" basis, relying upon the flexible space, labor, and contracting arrangements available in such areas. Uncertainty of demand precludes extensive advertising and attracts firms to their largest potential local markets.

With the stabilization of demand and with the standardization of the product and the methods of manufacture, firms in a maturing industry find it possible to pursue opportunities for cost reduction via large-scale production. At the same time their market horizon has broadened to encompass the whole nation. Being less dependent on the "incubator" advantages of the metropolitan area, they are left free to seek out locations which are attractive either in terms of geographic position with respect to a national market or in terms of the favorable cost structure they afford for critical factors of production.

In this framework, the fate of the Region as a location for industry is seen to depend upon the persistence of a high degree of uncertainty in the manufacture and marketing of certain products. The rapid growth of electronics has been accompanied by such uncertainty, first in radio, later in a more dilute form in television, and currently in the developing fields of military and industrial electronics.

The experience of the Region with the electronics industry presents a replica in miniature of what has happened to total manufacturing in the Region in the past quarter-century. The findings of a forthcoming study in this series suggest that the Region has just about maintained its share of the nation's manufacturing employment over this period. This surface stability, however, is seen as the resultant of two opposing forces: a persistent outward drift of certain industries in response to more favorable opportunities elsewhere, offset by the existence of rapidly growing industries in the Region's total industrial mix. This is very much the story of the electronics industry in the Region.

Future national demand and employment

We now have the difficult task of projecting the growth of the Region's electronics industry into the future. This has to be done in two steps. First, we need to project the growth of the industry in the nation as a whole. Then we must consider the extent to which the Region will participate in this growth. The former is by far the more hazardous and, unfortunately, the more critical step.

If electronics were solely entertainment products, the task of projecting would still be more difficult than it is in other industries. But with one-half or more of the industry's production going into national defense, we are confronted with the additional problem of anticipating the defense expenditures, influenced importantly by the course of international relations. The assumption which underlies our projection is that the burden of defense expenditures will remain fairly constant, at roughly 9 per cent of Gross National Product.

How much of total defense expenditures will go for electronics? In the last ten years electronics has taken a larger and larger share of the total. Between 1947 and 1957, defense expenditures in the aggregate increased by two and one-half times while defense expenditures for electronics increased ten times. The likelihood is that electronics will continue to take larger bites out of the defense budget. For one thing, electronics plays a bigger role in missiles than it does in aircraft, and missile procurement is increasing relative to the procurement of aircraft. In 1954, missiles accounted for only 10 per cent of the aircraft-plus-missiles defense budget. By 1958, however, the figure had risen to about 35 per cent and by 1960 it promises to reach 50 per cent. As a result *electronics* expenditure for missiles in the fiscal year 1957 was already higher than such expenditure for military aircraft.[2] This shift, coupled with the fact that even aircraft are being increasingly electronified, supports the expectation of a rapid rise in military electronics expenditure in the next decade, a rise more rapid than that for defense expenditures as a whole; beyond that time, however, we assume that military electronics will not grow faster than defense expenditures.

An even faster rate of growth is anticipated in the field of industrial electronics. The proliferating use of electronic computers in business, engineering, and governmental offices constitutes a major change in industrial practice. Sales of electronic computers to business and government, though only $350 million in 1957, could readily increase five-fold or six-fold by 1965.[3] And the same sort of expansion is expected for industrial electronics as a whole. However, the rate of growth cannot be expected to be equally rapid in the ensuing decade, and by the 1975–1985 decade it should no more than match the growth of the national economy.

Consumer products present a different story. Expenditure on such products has been rather stable since the initial flourish of television in the 1947–1950 period. Sales of such products have failed to expand since 1950 despite a rapid rise in national income. There are many reasons, however, why the level of expenditures can be expected to exhibit a steady upward trend in the future. With respect to TV particularly, one can anticipate a rising rate of replacement, a rise in the number of homes with sets, and a rise in the number of multiple-set homes. The impact which cheaper color television would have is as yet unknown. One cannot anticipate, however, that color TV will generate a boom as great as that which accompanied the introduction of black and white TV. More likely, it will gradually replace black and white TV production. Finally, the growth in high-fidelity reproduction, spurred presently by the new development known as stereophonics, can be expected to augment consumer sales.

It is obvious that some of these speculations are crude at best, and it would be foolish to project the growth of the electronics industry solely on the foundation which they offer. They are introduced here largely to suggest the type of thinking which underlies the dollar volume forecasts shown in Table 11. As can be seen, the industry

Table 11 Actual and Projected National Sales for the Principal Categories of the Electronics Industry in Selected Years (billions of dollars)

	Current dollars			1957 dollars		
Industry category	1947	1954	1957	1965	1975	1985
Electronics industry, total	1.2	5.1	7.6	14.0	20.8	29.8
Consumer electronics	0.8	1.4	1.5	1.9	3.0	4.0
Components for replacement	0.1 [a]	0.6	0.9	1.1	1.5	2.0
Military electronics	0.1 [a]	2.4	3.9	7.6	11.1	16.2
Industrial electronics	0.2 [a]	0.6	1.3	3.4	5.2	7.6

[a] Estimated.

Source: For 1947, 1954, and 1957, see *1958 Electronics Industry Fact Book* (Electronic Industries Association, Washington, D.C., 1958). For basis of projections, see Appendix E.

is expected to have grown nearly six-fold in the 31 years from 1954 to 1985. It is expected to double in the twenty years from 1965 to 1985.

Though projections of dollar volume are necessary and interesting, a more useful statistic for locational analysis is the number of jobs a given level of sales will provide, since it is primarily through payrolls that the peculiar regional impact of any industry is manifested. In 1954 there were 410,000 employees in the three Census industries which embraced most of the electronics industry. For purposes of projection we need to include a fourth industry, namely "Computing and Related Machines," because of the tremendous potential of electronic computers. Since that industry had 57,000 employees in 1954, the sum of the four industries was 467,000.

The methods by which we derived projections of future levels of employment in electronics are described in Appendix E. The critical assumptions, for which some support is given in the appendix, are that 40 per cent of the sales dollar will be devoted to payroll costs and that sales per worker, measured in constant dollars, will increase by 2½ per cent annually. The resultant projections of employment are shown in Table 12.

Table 12 Actual and Projected United States Employment in Electronics, 1954, 1965, 1975, and 1985

Year	Sales (billions)	Index of productivity (1954 = 100)	Employment
1954	$ 5.1	100	467,000
1965	14.0 [a]	131	1,000,000
1975	20.8 [a]	168	1,150,000
1985	29.8 [a]	215	1,295,000

[a] 1957 dollars.

FUTURE DEMAND AND EMPLOYMENT IN THE REGION

We now have to consider how the New York Metropolitan Region will participate in this anticipated growth. If our industry projections prove to be wide of the mark we shall of course have erroneous projections for the Region. But we can console ourselves with the anticipation that a projection of the relative position of the Region is probably subject to smaller error.

Again, the task is logically divisible into two parts. First we need to analyze the growth of the industry to see whether the Region with its peculiar contributions to electronics manufacturing will be favorably or adversely affected by changes in the product mix. Then, we need to consider whether the fundamental trends in the competitive position of the Region which were summarized at the outset of this chapter should be assumed to persist into the future.

The shift in products. More than once we have emphasized the point that the only certain thing about electronic products is the fact that they will change. Does this mean that one can envisage a constant flow of new products, each incubating in the New York Metropolitan Region because of its demonstrated affinity for such products, then spreading out to the other centers? We can test this possibility in a dilute way by examining the locational implications of some typical new products which are now moving into quantity production.

The transistor is one such product. This device, the size of a pencil eraser, regulates the flow of electrons as does its ancestor, the vacuum tube. Yet it has many advantages over the vacuum tube. It occupies much less space, is more durable, weighs less, and requires less power to operate. So far, the main factor that has inhibited its replacement of the vacuum tube has been cost.

The odds are very high, however, that these cost differences will be overcome in the next five or ten years. Sylvania Electric Products, Inc., has estimated, for instance, that by 1966 from 60 to 80 per cent of requirements for radio tubes and for receiving tubes in black and white TV sets will be met by transistors. Even if the price handicap is not overcome, the growth of the market for computers—a product which will probably use transistors much more than tubes—should add substantially to transistor sales.

Though the major producers of electronic tubes were once located in the New York Metropolitan Region, the largest producer of the tube's rival, the transistor, is Texas Instruments in Dallas, Texas. The "old line" tube-producing companies, like General Electric and Sylvania, are not making transistors in the New York Region. Transistor production in the New York area is largely confined to specialized types such as those made by General Transistor, Amperex, and Bendix Aviation. The only manufacturers in the area which have mass-production experience in transistors are Tung-Sol in Newark and RCA at Somerville, New Jersey. Accordingly, the present disposition of transistor facilities does not lead one to predict that the New York area will be in the forefront as the industry moves from tubes to transistors over the next decade.

Color television must also be considered. It is doubtful that a new wave of additional facilities will be required to meet the growing demand for this product. It is not a new medium, but only an improvement on an existing one. Neither its technology nor its demand will be subject to an initial period of frenetic uncertainty. When that demand does develop, it will be national in scope, not initially centered on the eastern seaboard. Consequently, the attractions which the New York Metropolitan Region might ordinarily exercise for a new industry simply will not apply in this case.

High-fidelity sound, however, suggests another pattern. In the chapter on consumer electronics we emphasized the distinction between the production of (1) the mass-produced items of radio, TV, and phonographs and (2) hi-fi sub-assemblies and expensive, high-quality packaged items. Within the past few years a change of the public's tastes has made the assembly of high-quality sound reproduction equipment a bonanza, constituting the difference between profits and losses for some firms. There has been not only a shift to serious music by the listening public, but also a shift away from the fare presented both by AM radio and TV. Frequency modulation radio has enjoyed a rebirth along with the high-fidelity phonograph, and FM, it is asserted, "is prospering in direct ratio as it provides adults with a refuge from the blaring Children's Hour of the AM juke boxes and from the vacuity of most TV."[4]

Currently a new dynamic element is being injected into the hi-fi picture with the development of stereophonic sound, which adds a sense of "depth" analogous to the improvement in motion pictures introduced by "Cinerama" and similar techniques. The unsettled state of technology in this field could be expected to favor the New York Metropolitan Region in the short run. Until technology has stabilized, no manufacturer will wish to make large commitments to mass production. For example, RCA has been working on a method of compatible stereophonic broadcasting for AM radio. The longer this period of gestation, the longer will the Region continue as a major factor in the production of high-quality sound reproduction equipment. To be sure, packaged stereophonic phonographs are being advertised by manufacturers in various parts of the country, but at this writing it would be premature to suppose that stereophonics has already come of age. In any case the packaging of stereophonics will eventually follow the path already trod by radio, television, and packaged hi-fi equipment, though still leaving in the New York area custom equipment and the high-quality sub-assemblies which delight the audiophile.

What the pattern seems to add up to is something along the following lines. New products will constantly develop, supplementing and replacing the old. Some of these products will entail considerable uncertainty both in demand and in technology; others will be produced for a market which is more predictable or by a technology which is less volatile. This lessened uncertainty will stem partly from the fact that the new products will be replacements for others already in use, as the transistor is the replacement for the tube, and partly from the industry's growing ability to project the market for new electronic devices. On the whole, therefore, it seems inevitable that the level of uncertainty associated with these new products should be rather less than that associated with the earlier electronic products. Accordingly, the elements of the environment of the New York Metropolitan Region which make it especially attractive for new products in general should be a less compelling force in the location of the new electronic products of the future than it has been in the past.

The shift in costs. Quite apart from the change in products, there is also the possibility that freight, labor, and other costs also could operate to alter the industry's locational pattern. For one thing, we expect no easing of the pressure to locate set-assembly plants closer to the faster-growing markets of the nation. There is every reason to believe that the long-run trends in the geographic redistribution of population and purchasing power will continue. First, the western states will continue to increase their share of the nation's population. Second, income in the areas of low per capita income will continue to increase more rapidly than in the areas of high per capita income. Taken together, these trends spell a further decline in the share of the national market located in the Northeast.

Likewise we expect no significant improvement in the Region's labor-cost position—none, at any rate, of sufficient magnitude to withstand the attraction that areas of lower-cost labor exert on the production of standardized components. On the other hand there is likely to be only a modest decline in the Region's share of military electronics, since this is one branch of the industry whose technical requirements and demand characteristics are least likely to settle down. The emphasis, according to spokesmen for the Air Materiel Command, is on a "shift from high volume or mass production to precision production of more limited quantities of increasingly complex material." [5]

Yet even in military electronics, the mainstay of such an area as the New York Metropolitan Region, the tide must plainly begin to run at some stage in favor of other areas. The New York Region began with a lead which only a few other areas, such as New England, could rival. Apart from such advantages as rentable space, testing laboratories, and extensive university facilities, the New York Region could also provide a floating supply of engineers and a variety of skilled subcontractors.

The newer areas of military production, such as Florida, California, and Colorado, are already competing in the type of production which can be organized on a self-contained basis, without recourse to the external economies of their respective areas. The presence of a large number of electronics firms in these competing areas, whatever the character of those firms, will eventually breed the external economies which New England and the New York area have long possessed. The fragmented facilities, the specialized subcontractors, the floating pool of engineers—all of these attributes are gradually being produced in competing markets. Eventually, this process should whittle down the initial lead of the older areas over the new.

The Region's share. It is a seven-league-boot leap from the kind of qualitative considerations which have been discussed in the preceding sections to an appraisal of the Region's future share of the various categories of electronics production. In making the leap, however, one powerful force—inertia—is on the side of the projector. Even in industries whose products and total levels of activity are shifting dramatically, the share of a large area like the New York Metropolitan Region tends to change sedately.

Table 13 reflects our judgment of the speed with which the New

Table 13 Sales of the Principal Categories of the Electronics Industry in the New York Metropolitan Region, Estimated for 1954 and Projected for 1965

	Sales, 1954		Sales, 1965	
Industry category	Millions of dollars	Per cent of U.S. sales	Millions of dollars	Per cent of U.S. sales
Electronics industry, total	896	17.6	2,430	17.4
Consumer electronics	180	13	190	10
Components for replacement ..	66	10	77	7
Military electronics	500	20	1,368	18
Industrial electronics	150	25	795	23

York Metropolitan Region's share of production will change in various categories of the electronics industry. The 1954 breakdown in the table will not be found in any source hitherto published, since it is based on confidential estimates made within the industry. The 1965 proportions for the New York Metropolitan Region, by categories, are our quantitative embodiment of the qualitative factors heretofore discussed.

For the years beyond 1965, our courage fails in projecting the shares for individual categories. At this stage, we are prepared to foresee a continued relative decline for the Region's aggregate electronics production, though at a lesser rate. Against a Region share of 17.6 per cent in 1954, and 16.3 per cent in 1965, we suggest a ratio of 17.0 per cent in 1975 and 16.5 per cent in 1985.

What would this volume of sales over the years mean in terms of employment? Table 12 had suggested that national employment in the four branches of the industry which come closest to representing our future electronics "universe" would be 1,000,000 in 1965;

1,150,000 in 1975; and 1,260,000 in 1985. If we apply the percentage shares just suggested to these figures, an estimate of the New York Metropolitan Region's employment emerges. A declining share offset by a rising national total produces major increases in employment for the Region, as follows:

1954	87,000
1965	175,000
1975	195,000
1985	215,000

Meanwhile, significant shifts will go on in the location of electronics employment *within* the Region. The Region as a whole will continue to specialize in a swiftly changing product, with small, uncertain firms continuing to play a major part in the total industrial make-up. With research and development so significant an element in the picture, the disposition to congregate in close proximity to other establishments will continue strong; so will the disposition to occupy rented quarters on flexible terms.

Yet some outward redistribution of total employment in these industries seems inevitable within the Region. The increasing industrial densities in suburban areas like western Suffolk County and western Essex County will gradually provide the rentable space, the floating supply of engineers, the testing laboratories, and the subcontractors which heretofore have been most prominently available in New York City and the older portions of Essex County. With the external economies of the older areas matched by those of the suburbs, an outward drift will almost certainly result.

Despite the operation of such forces, New York City managed to increase its absolute level of employment between 1947 and 1956. Yet there are strong signs pointing to an absolute decline in the future. In 1947 there were still some undeveloped industrial tracts in Queens and Brooklyn. These are being used up at a pace which suggests that there will be hardly any room for expansion on new sites in the near future. At the same time, the available loft space is diminishing as the pressure of alternative demands generates redevelopment along entirely different lines. Also, the larger firms in the Region already located outside New York City and engaged in military and industrial work will contribute to the more rapid growth of the Region's outlying counties as a consequence of the more rapid growth of these segments of the electronics industry.

As we saw in Table 10, New York City's share of the Region's employment in the group of three major electronics industries declined from 36 to 24 per cent between 1947 and 1956. When we include the fourth industry, "Computing and Related Machines," the decline was from 31 to 21 per cent. Table 14, which covers all

Table 14 Actual and Projected Employment in Electronics Industries in Various Parts of the New York Metropolitan Region

	Thousands of employees				Percentage of NYMR employment			
	1956	1965	1975	1985	1956	1965	1975	1985
NYMR, total	113	175	195	215	100	100	100	100
New York City	23	30	27	26	21	17	14	12
Nassau and Suffolk	21	33	39	45	18	19	20	21
Other New York counties and Fairfield	16	28	33	39	14	16	17	18
New Jersey counties	53	84	96	105	47	48	49	49

Source of 1956 data: Unpublished figures of state departments of labor as collected under unemployment insurance programs, 3rd quarter, 1956, except for Fairfield County employment, which is for 1st quarter.

four industries, shows our projections for the various parts of the Region. We anticipate that New York City's electronics employment will reach its peak in absolute terms by 1965, at which time its share of the Region's total will have further declined from 21 to 17 per cent. All other sections of the Region will increase their shares as a consequence of a further outward drift from the city, though the increases will be at a reduced rate. The total in the group called "Other New York counties and Fairfield" should be buoyed by the rapid growth of computer production in Dutchess County.

We may take it as certain that there will be developments in electronics between now and 1985 which are unpredictable today—developments which may reach to the stars, and may or may not upset the employment forecasts for the New York Metropolitan Region. Frank W. Mansfield, dean of market analysts for the electronics industry, a man who stands on the corner where economics and electronics meet, recently challenged the claim that the electronics industry has been "over-glamorized." He said, "The future of electronics and its applications is breath-taking, even if we consider only what now appears certain for the years ahead. From our present knowledge of the potential of electronics, it appears that no fantastic story we could possibly make up will surpass what lies over the horizon." [6]

APPENDICES
AND
NOTES

Women's and Children's Apparel

APPENDIX A

CLASSIFICATION OF APPAREL INDUSTRIES

The purpose of this appendix is to give fuller descriptions of the women's and children's apparel industries than could be given in the text without insufferably retarding the reader.

The apparel study is concerned with twelve so-called "four-digit" industries as defined in the Standard Industrial Classification (S.I.C.) which is laid down by the Executive Office of the President, Bureau of the Budget. For example, one of them is 2331 (blouses) and another is 2333 (dresses, unit-priced). Groups of these "four-digit" industries make up "three-digit" industries; to illustrate, six of them constitute the women's and misses' outerwear industry which bears the number 233.

All the industries covered in our report are listed in Table 1, page 25, but not with their official numbers and not always with the technical titles assigned to them by the government. They are listed here in greater detail.

233. *Women's and misses' outerwear.* This "three-digit" group makes up the most important segment of the women's and children's apparel industries and includes blouses, dresses, coats, suits, skirts, and other outer garments. The constituent industries within women's and misses' outerwear are:

2331. *Blouses* (full S.I.C. title: "Women's and Misses' Blouses and Waists"). This industry comprises establishments primarily engaged in manufacturing women's and misses' blouses, including nonknitted shirts.

2333. *Dresses, unit-priced* (full S.I.C. title: "Women's and Misses' Dresses"). This industry comprises establishments primarily engaged in manufacturing women's and misses' dresses. Its products are usually sold by the piece, and are designed for street wear. This dress industry is the largest single component of women's and children's apparel.

2334. *Dresses, dozen-priced* (full S.I.C. title: "Household Apparel"). This industry comprises establishments primarily engaged in the production of household apparel, chiefly of washable fabrics, including aprons, smocks, hoovers, house dresses, and nurses' and maids' uniforms. Its products are usually sold by the dozen.

2337. *Coats, suits, skirts* (full S.I.C. title: "Women's and Misses' Suits, Coats [Except Fur Coats], and Skirts"). This industry comprises establishments primarily engaged in manufacturing women's and misses' suits, ensemble suits, skirts, and coats other than fur. It is the second largest component of women's and children's apparel. In the *1947 Census of Manufactures,* the statistics for Industry 2337 were published as two separate sub-industries, "Women's Suits and Coats," and "Women's Skirts."

2338. *Neckwear* (full S.I.C. title: "Women's Neckwear and Scarfs"). This very small industry comprises establishments primarily engaged in manufacturing women's neckwear, scarfs, and dickies, cut and sewed from woven or knit fabric.

2339. *Other outerwear* (full S.I.C. title: "Women's and Misses' Outerwear, Not Elsewhere Classified"). This industry comprises establishments primarily engaged in manufacturing other women's outerwear, including bathing suits, beachwear, swimwear, slacks, riding habits, sweaters and outerwear sport shirts, cut and sewed from purchased woven or knit fabric.

234. *Women's and children's undergarments* (full S.I.C. title: "Women's, Misses', Children's and Infants' Undergarments"). The constituent industries of this "three-digit" industry are:

2341. *Underwear, nightwear* (full S.I.C. title: "Women's, Misses', Children's and Infants' Underwear and Nightwear"). This industry comprises establishments primarily engaged in the manufacture of women's, misses', children's and infants' underwear and nightwear, cut and sewed from purchased woven or knit fabric. Its primary products include slips, panties, nightgowns, sleeping pajamas, and negligees.

2342. *Corsets, brassieres* (full S.I.C. title: "Corsets and Allied Garments"). This industry comprises establishments primarily engaged in manufacturing corsets and corset accessories, brassieres, girdles, foundation garments, and similar body-supporting garments.

236. *Children's outerwear* (full S.I.C. title: "Children's and Infants' Outerwear"). The constituent industries of this "three-digit" industry are:

2361. *Dresses* (full S.I.C. title: "Children's and Infants' Dresses"). This industry comprises establishments primarily engaged in manufacturing children's and infants' dresses, blouses, blousettes, waists and shirts, cut and sewed from purchased woven or knit fabric.

2363. *Coats* (full S.I.C. title: "Children's and Infants' Coats"). This industry comprises establishments primarily engaged in manufacturing children's and infants' coats, coat-and-legging sets, and snow suits, cut and sewed from purchased woven or knit fabric. Establishments engaged in manufacturing these garments in teen-age girls' and girls' sizes are also included.

2369. *Other outerwear* (full S.I.C. title: "Children's and Infants' Outerwear, Not Elsewhere Classified"). This industry comprises establishments primarily engaged in manufacturing other children's outerwear, including house coats, middies, slacks, beachwear, rompers, sun suits, play suits, bathrobes, headwear, and skirts, cut and sewed from purchased woven or knit fabric.

The following is officially classified as a textile industry but is included in our study for the reasons given on page 23:

2253. *Knit outerwear* (full S.I.C. title: "Knit Outerwear Mills"). This industry comprises establishments primarily engaged in knitting outerwear, or in manufacturing outerwear from knit fabric produced in the same establishment. Its important products include sweaters, bathing suits, and dresses. It also includes some men's wear garments.

APPENDIX B

ADDITIONAL APPAREL STATISTICS

Table B–1 Employment in Six Major Apparel Branches, by Counties of the New York Metropolitan Region, 1956[a]

	Dresses, unit-priced	Coats, suits, skirts	Rest of women's outer-wear[b]	Under-garments	Chil-dren's outer-wear	Knit outer-wear
Entire Region	78,765	57,402	27,158	41,682	30,510	24,286
New York City[c]	64,420	41,578	21,206	29,514	21,220	18,259
Outside New York City, total	14,345	15,824	5,952	12,168	9,290	6,027
Nassau	522	721	359	91	268	188
Suffolk	769	782	229	158	547	256
Westchester-Rockland	2,963	515	635	334	451	470
Dutchess-Putnam-Orange	765	1,498	393	988	344	d
Hudson	1,779	3,376	730	3,185	1,979	2,635
Essex	1,636	2,191	541	1,096	1,200	657
Passaic	624	3,372	483	1,067	1,382	584
Bergen	1,510	1,111	146	569	506	457
Union-Somerset-Morris	891	375	598	1,026	798	236
Middlesex	1,462	459	872	878	969	124
Monmouth	476	935	284	481	679	d
Fairfield	948	489	682	2,295	167	324

[a] Average of July and September, the only two months for which adequate data are available. However, Fairfield County employment is for first quarter. Employees included in this table are those covered by state unemployment insurance programs—slightly less than total employment.

[b] Blouses, dozen-priced dresses, neckwear, and women's "other outerwear."

[c] Separate boroughs not available.

[d] Data not revealed.

Source: Unpublished figures from departments of labor of New York State, New Jersey, and Connecticut.

NOTES TO WOMEN'S AND CHILDREN'S APPAREL

CHAPTER 1: THE BUSINESS OF GARMENT-MAKING

1. The sales percentage based upon data of the National Credit Office, Inc.

2. *The Womenswear Industry* (New York: Market Planning Service, National Credit Office, 1950), pp. 18, 49.

3. Data of the United States Treasury Department concerning 1953 Corporation Income Tax Returns show that the compiled net profits of the firms in the women's clothing industry were $49,275,000; their gross sales amounted to $3,681,544,000; the net profits as a percentage of sales therefore were only 1.34. In that year, there were more returns with no net income than with net income. Those reporting net income had $2,908,738,000 worth of gross sales and a compiled net profit of $79,276,000, a ratio of 2.73 per cent. Those reporting no net income for 1953 had gross sales of $772,806,000 and a compiled net loss of $30,001,000, a ratio of 3.88 per cent.

4. *The Womenswear Industry,* pp. 14, 31, 45.

5. U.S. *1954 Census of Manufactures.*

6. *Business Week,* Aug. 18, 1956, p. 132.

7. U.S. Bureau of Labor Statistics, *Case Study Data on Productivity and Factory Performance . . . Women's Dresses* (Washington, 1955), p. 1.

8. Margaret Wray, "Fashion in the Women's Outerwear Industry," *Westminster Bank Review,* London (November 1954), p. 7.

9. Abe S. Weiss, *Garment Workers of Other Lands* (New York: Educational Department, International Ladies' Garment Workers' Union, mimeo., circa 1940), p. 7.

10. *Eighth Census of the United States, 1860,* as quoted in Lewis Lorwin, *The Women's Garment Workers* (New York: B. W. Huebsch, 1924), p. 5. (This book bears the name Louis Levine, but the author later changed his name to Lewis Lorwin.)

11. U.S. Bureau of Labor Statistics (see note 7, above), p. 1.

12. Solomon Fabricant, *Employment in Manufacturing, 1899–1939* (New York: National Bureau of Economic Research, 1942), p. 48.

13. Alfred E. Parrott, *The Future of the New York Apparel Industry, 1952–1970 (Preliminary)* (New York: Regional Plan Association, 1953), pp. 24, 25.

14. Computed from data of the New York State Department of Labor, published in its monthly series, *Labor Market Review.*

15. Gertrude Berta Greig, *Seasonal Fluctuations in Employment in the Women's Clothing Industry in New York* (New York: Columbia University Press, 1949), p. 62.

16. Greig, p. 55.

17. Computed from data of the New York State Department of Labor, ap-

pearing in its monthly series, *Labor Market Review,* and from data of the U.S. Bureau of Labor Statistics, appearing in its *Monthly Labor Review.*

18. Charles S. Goodman, *The Location of Fashion Industries, with Special Reference to the California Apparel Market,* Vol. 10, No. 2 of *Michigan Business Studies* (Ann Arbor: University of Michigan Press, 1948), p. 1.

19. Victor R. Fuchs, *The Economics of the Fur Industry* (New York: Columbia University Press, 1957), p. 57.

20. The research director of the union, testifying before a Congressional committee, claimed that the mergers in the textile industry were overbalancing the textile firm's power vis-a-vis the apparel manufacturer, and that garment manufacturers were being forced to purchase far in advance of their needs. He contended also that the rise of the department stores, mail order houses, and resident buying offices had put the retailers in a position to exploit the garment manufacturer by insisting on special concessions. *Trends and Prospects, Women's Garment Industry, 1947–1950* (New York, International Ladies' Garment Workers' Union), pp. 22–24.

CHAPTER 2: NEW YORK'S DOMINANCE

1. Hamilton's "Report on the Subject of Manufactures" (1791), in *Industrial and Commercial Correspondence of Alexander Hamilton,* ed. Arthur Harrison Cole (Chicago, 1928), p. 280.

2. This paragraph based on Paul H. Nystrom, *Economics of Fashion* (New York: Ronald Press, 1928), pp. 404–411.

3. Lewis Lorwin, *The Women's Garment Workers* (New York: B. W. Huebsch, 1924), p. 12. (This book bears the name Louis Levine, but the author later changed his name to Lewis Lorwin.)

4. Lorwin, pp. 14, 15.

5. Jesse Eliphalet Pope, *The Clothing Industry in New York* (University of Missouri, 1905), pp. 42, 43.

6. Lorwin, p. 18.

7. Maurice Dobb, *Wages* (Cambridge, Eng.: Cambridge University Press, 1955), p. 77.

8. Nystrom, p. 412.

9. Lorwin, p. 100.

10. Lorwin, p. 384.

11. Edward Ewing Pratt, *Industrial Causes of Congestion of Population in New York City* (New York: Columbia University Press, 1911), pp. 80, 81.

12. B. M. Selekman, Henriette R. Walter, W. J. Couper, *The Clothing and Textile Industries,* Vol. 1B of *Regional Survey of New York and Its Environs* (New York: Regional Plan of New York and Its Environs, 1928), Table XXI.

13. S. P. Dobbs, *Clothing Workers of Great Britain* (London: Rutledge, 1928), p. 69.

14. Charles S. Goodman, *The Location of Fashion Industries, with Special*

Reference to the California Apparel Market, Vol. 10, No. 2 of *Michigan Business Studies* (Ann Arbor: University of Michigan Press, 1948), pp. 11, 12.

15. *New York Times,* Feb. 9, 1958, reporting a market study by the Olen Company, a chain of 119 stores.

16. *Women's Wear Daily,* March 11, 1958.

17. Market Planning Service, National Credit Office, *Markets, Fabrics, and Garments of the Apparel Manufacturing Industry, 1955–1956,* pp. 6–14. The number of *firms* should not be confused with the much larger number of *establishments* given in Chapter 1.

18. *Women's Wear Daily,* March 24, 1958.

19. Firms and their addresses obtained from the respective industrial directories issued by Fairchild Publications in 1958.

20. *Report and Record, Twenty-Ninth Convention, International Ladies' Garment Workers' Union, 1956,* p. 170.

21. John I. Griffin, *Industrial Location in the New York Area* (New York: The City College Press, 1956), p. 52.

CHAPTER 3: NEW YORK'S DECLINE AS A PRODUCTION CENTER

1. Leonard A. Drake and Carrie Glasser, *Trends in the New York Clothing Industry* (New York: Institute of Public Administration, 1942), pp. 51, 52. The "women's clothing industry," as here defined, is approximately the same as the twelve industries covered by our study and defined in Appendix A, except that it omits the corset-and-brassiere industry and knit outerwear. The term "wage earners" is about the same as production workers.

2. U.S. *1947 Census of Manufactures.* The figures from that census are not strictly comparable with those of 1939, but they do show that the number of production workers in the twelve women's and children's apparel industries under consideration (except knit outerwear) in the boroughs of Manhattan, Brooklyn, the Bronx, and Queens in 1947 was still 40.1 per cent of the number in the nation. The 48.7 per cent figure for the Standard Metropolitan Area is computed on the basis of all twelve industries and total employment (production and nonproduction workers). See Table 6.

3. The metropolitan area's relative growth in the corset-and-brassiere industry is misleading, because the totals for the United States do not include Puerto Rican production. There are no published data on the size of the labor force in the corset-and-brassiere industry of Puerto Rico in 1947, but it must have been tiny because even in 1949 there were only two manufacturing establishments. But according to the *Puerto Rico Census of Manufactures, 1954,* by that year the employment in the corset-and-brassiere industry had reached 2,953, which, if added to that of the 48 states, would make the proportional growth in the United States greater than that in the New York area.

4. These percentages, and the others in this and the next paragraph, are from the U.S. *Census of Manufactures* for the years indicated.

5. "South" here includes three Census regions: South Atlantic, East South Central, and West South Central.

6. See Fessenden S. Blanchard, "The Revolution in Clothes," *Harper's*, March 1953.

7. Figures computed from U.S. Bureau of the Census, *Facts for Industry* series.

8. ILGWU, *Trends and Prospects, Women's Garment Industry, 1953–1956*, pp. 6, 7.

9. Nathan Belfer, "Section Work in the Women's Garment Industry," *Southern Economic Journal*, 21:189 (Chapel Hill, October 1954).

10. Isidore Nagler, *Analysis of the Problems of the Women's Coat and Suit Industry and Suggested Recommendations for Their Solution*, mimeo., 1954.

11. U.S. Bureau of Labor Statistics, *Occupational Earnings, Women's and Misses' Dresses, August 1955*.

12. Women's Bureau, U.S. Department of Labor, Bulletin 141, *Piecework in the Silk Dress Industry* (Washington, 1936).

13. William L. Batt, Jr., Pennsylvania Secretary of Labor and Industry, "Pennsylvania's Labor Surplus," in *Pennsylvania*, special section of the *New York Times*, Nov. 3, 1957, p. 30.

14. Pennsylvania Bureau of Employment and Unemployment Compensation, *The Women's Outerwear Industry in Pennsylvania*, Aug. 3, 1951, p. 5.

15. Glenn E. McLaughlin and Stefan Robock, *Why Industry Moves South* (National Planning Association's Committee of the South, 1949), pp. 75, 76.

16. Seymour E. Harris, *The Economics of New England* (Cambridge: Harvard University Press, 1952), p. 137.

17. *The Pressure's On—Will the Garment Manufacturers Yield to the Beckoning of Nearby Communities?*, ILGWU Research Department memorandum, circa 1945.

18. Percentages in this paragraph were computed by Martin Segal and will be included in his forthcoming volume (a part of this series) on labor as a locational force.

19. *Journal of Commerce*, New York, March 6, 1923.

20. ILGWU, *Trends and Prospects, Women's Garment Industry, 1937–1940*, p. 2.

21. ILGWU, *Trends and Prospects, Women's Garment Industry, 1953–1956*, p. 13.

22. Computed from data of the ILGWU Research Dept., reporting the following changes between 1939 and 1955: corset-and-brassiere production workers increased from 18,800 to 36,100 and union membership from 4,722 to 16,967; knit outerwear production workers increased from 23,400 to 42,900 and union membership from 6,890 to 21,248.

23. Leo Troy, *Distribution of Union Membership among the States, 1939 and 1953* (New York: National Bureau of Economic Research, Occasional Paper 56, 1957), p. 12.

24. U.S. Bureau of Labor Statistics, *Case Study Data on Productivity and Factory Performance . . . Women's Dresses* (Washington, 1955), p. 81.

25. Julius Hochman, "More About Price Settlements" (reprinted from *Justice,* ILGWU publication, Sept. 1 and 15, Oct. 1, 1956), p. 7. (The italics are Hochman's.)

26. U.S. Bureau of the Census, Report No. FT 110, for 1954, 1955, 1956, 1957.

27. Will Herberg, "The Old-Timers and the Newcomers," *The Journal of Social Issues,* 9:13 (1953).

28. Mabel Durham, *Economic Trends in the Women's Undergarment Industry* (typewritten, circa 1946), p. 6.

29. See the author's *Puerto Rican Integration in the Skirt Industry in New York City,* mimeo. (New York State Commission Against Discrimination, 1958).

30. Herberg, p. 15.

31. U.S. Bureau of Labor Statistics, Report No. 122, *Wage Structure, Women's and Misses' Coats and Suits, February 1957* (Washington, November 1957), p. 9.

32. *New York Times,* March 27, 1958.

33. Compare "Puerto Rican Migration to New York City," *Bulletin,* Department of City Planning, City of New York, February 1957.

34. *A Summary in Facts and Figures, Progress in Puerto Rico,* Puerto Rican Migration (New York: Commonwealth of Puerto Rico, April 1957), p. 16.

35. "Puerto Rican Migration to New York City," p. 5.

36. "Gary Meets Recession in a Business-as-Usual Mood," *Business Week,* May 24, 1958.

37. Research Department, New York Cloak Joint Board, ILGWU, *Survey of Shops and Workers in New York Coat-and-Suit Industry, Spring 1953.* Shops in South Jersey are excluded from this breakdown, because they are not in the New York Metropolitan Region.

38. *New York Times,* June 13, 1958.

39. "Commercial and Industrial Floor Space Inventory," *Bulletin,* Department of City Planning, New York City, December 1957.

40. *New York Times,* March 31, 1958.

41. See note 39.

42. *New York Times,* Nov. 7, 1958.

43. Alfred G. Dale, "Texas' Expansive Clothing Industry," *Texas Business Review,* November 1954.

44. Based upon estimates of employers interviewed. Direct comparisons between rents in New York and elsewhere are complicated by the fact that New York rents include services, whereas in other locations, the tenant must provide his own heat and maintenance and pay water taxes, etc. A typical advertisement in *Women's Wear Daily,* April 9, 1958, offered a garment plant in

the hard-coal area of Pennsylvania at an annual rental of 30 cents per square foot; with heat and maintenance added, the total rent would be close to the 50 cents per square foot estimated by the employers interviewed.

45. *A Stitch in Time,* Mayor's Committee for World Fashion Center, New York, 1957.

46. The methodology of arriving at this cost estimate is explained in another volume of this series, Edgar M. Hoover and Raymond Vernon, *Anatomy of a Metropolis* (Cambridge: Harvard University Press, 1959), Appendix E.

47. For a discussion of the traffic situation see letter of the Mayor's Interdepartmental Traffic Council to the members of the New York City Board of Estimate, Jan. 16, 1958. The 475 estimate is from *A Stitch in Time.*

48. New York City Department of Commerce and Public Events, *Highlights,* July–August 1957, p. 7.

49. *Business Week,* Aug. 18, 1956.

50. Arthur L. Reuter, Acting Commissioner of Investigation of the State of New York, *Report on the Activities and Associations of Persons Identified as Present . . . at Apalachin, New York, on November 14, 1957, and the Reasons for Their Presence,* mimeo., April 23, 1958, p. 28.

51. *Women's Wear Daily,* Feb. 21, 1958.

52. Glenn E. McLaughlin and Stefan Robock, *Why Industry Moves South* (National Planning Association's Committee of the South, 1949), p. 114.

53. U.S. Bureau of Labor Statistics, *Case Study Data on Productivity and Factory Performance . . . Women's Dresses* (Washington, 1955). These payments were estimated to be 12 per cent, but another 0.5 per cent was added in 1958 for severance pay.

CHAPTER 4: THE FUTURE

1. In this case, and in our whole discussion of aggregate demand, "Women's and children's clothing" includes not only the products of the industries covered by our report, but also millinery, accessories, fur garments—in fact everything that women and children wear except footwear. The reason for the change of coverage is that the United States Department of Commerce uses this broader coverage in its estimates of personal consumption expenditures, upon which we depend here.

2. Real "per female" personal consumption expenditures for women's and children's clothing are computed by deflating the expenditures by the consumer price index for clothing and dividing by the female population of the nation. Real disposable income per capita has been computed by the U.S. Department of Commerce, *National Income,* 1954 edition, pp. 24, 25, for the years 1929–53—and by the writer for the subsequent years—by using the Department's "implicit deflator" applicable to personal consumption expenditures and dividing by the United States population.

3. The industry's garments include boys' coats in sizes 2 to 8.

4. Editors of Fortune, *The Changing American Market* (New York: Hanover Press, 1953), p. 186.

5. See *Summary of Family Expenditures for Clothing for Women and Girls and Children under 2 Years,* which is Volume 6 of *Study of Consumer Expenditures, Incomes and Savings,* tabulated by Bureau of Labor Statistics for Wharton School (University of Pennsylvania, 1956), pp. 10–20.

6. Granville Hicks, "How We Live Now in America," *Commentary,* December 1953.

7. J. Frederic Dewhurst and associates, *America's Needs and Resources, A New Survey* (New York: Twentieth Century Fund, 1955), p. 176.

8. "Sixty-Six Million More Americans" in *Readings in Economics from Fortune,* ed. Richard E. Mulcahy (New York: Henry Holt, 1957), p. 19.

9. *Summary of Family Expenditures* (our note 5), pp. 10–20.

10. *Summary of Family Expenditures,* p. 48.

11. Dewhurst, p. 176.

12. *Women's Wear Daily,* June 6, 1957, based upon data of the U.S. Department of Commerce.

13. Thomas M. Stanton, *Factors Determining Clothing Expenditures* (Wilmington: E. I. du Pont, mimeo., circa 1957), pp. 1, 14, 15.

14. Projection of the New York Metropolitan Region Study, based upon the assumption of 2½ per cent annual growth in productivity.

15. If the four years 1945 through 1948, in which demand for apparel was abnormal because of the shortage of durable goods, are omitted, the trend from 1929 to 1956 becomes: $Y = 5.150 - .055X$. Where Y is the percentage of disposable income spent on women's and children's clothing, X is the number of years from 1940, and the parameters are expressed in percentages.

16. This trend was calculated by the writer by using the method of least squares, yielding the equation: $Y = \$49.65 + .0571X$. Y represents expenditures per female expressed in 1947–49 clothing dollars, and X represents per capita disposable personal income in 1947 dollars. The Bureau of Labor Statistics consumer price index for apparel is used to convert the anticipated expenditures from 1947–49 clothing dollars into 1954 dollars.

17. Editors of *Fortune* (see note 4, above), p. 181.

18. See note 14.

19. See note 15.

Printing and Publishing

APPENDIX C

CLASSIFICATION OF PRINTING AND PUBLISHING INDUSTRIES

As an aid to the nontechnical reader, and to avoid cluttering up the text with minutiae, this appendix presents the classification scheme used by the Bureau of the Census in reporting figures on the printing and publishing industries.

The basic unit which the Census classifies is the establishment: that is, a plant at a separate location from others operated by the same company. An industry is defined as a group of establishments producing a single product or a group of more or less closely related products. In some instances the classification is based on the end product; establishments publishing books form one industry. In other cases, the classification is based on the process involved; commercial printing and lithographing are distinguished by the nature of the process rather than by the end products.

The group of products which is assigned to an industry is "primary" to that industry; an establishment is classified in a particular industry if its production of the primary products of that industry exceeds in value its production of the products of any other industry. Accordingly, if an establishment published 51 per cent books and 49 per cent periodicals, all of its employment would be shown in the book-publishing industry. Fortunately, such cases are rare in the printing and publishing industries (industry specialization is quite high for almost all of them), but this qualification must be borne in mind in interpreting Census statistics.

There follows a list of the printing and publishing industries, along with their Standard Industrial Classification (S.I.C.) code numbers:

2711. *Newspapers.* Composed of establishments that publish newspapers, or publish and print them. Establishments that print newspapers but do not publish them are classified in Industry 2751 or 2761.

2721. *Periodicals.* Composed of establishments that publish and print periodicals, or that publish only. If the establishment only *prints* periodicals, it is classified in Industry 2751 or 2761.

⁂

273. *Books.* This "three-digit" industry consists of two "four-digit" ones:

2731. *Books: Publishing and Printing.* Composed of establishments primarily engaged in publishing books and pamphlets, or in publishing and printing them.

2732. *Book Printing.* Composed of establishments primarily engaged in printing, or in printing and binding (but not publishing) books and pamphlets.

⁂

2741. *Miscellaneous Publishing.* Composed of establishments primarily engaged in publishing maps, atlases, sheet music, directories, racing forms, and other miscellaneous material not covered under the classifications "Periodicals" and "Books." Establishments primarily engaged in printing these materials, but not engaged in publishing, are classified in either Industry 2751 or 2761.

2751. *Commercial Printing.* Includes establishments primarily engaged in printing by letterpress, gravure and screen process. (The use of the gravure and screen processes accounts for a small fraction of the output of the industry.) In particular, shops which specialize in printing newspapers or periodicals for others are included within this industry (unless their principal process is lithography). Shops which specialize in printing books for others, however, are included in Industry 2732.

2761. *Lithographing.* Composed of establishments primarily engaged in preparing lithographic stones or plates and in printing from them. Lithographing is the process usually known as "offset."

2771. *Greeting Cards.* Composed of establishments primarily engaged in designing only, manufacturing only, or both designing and manufacturing greeting cards.

⁂

278. *Bookbinding and Related Industries.* Consists of four sub-industries:

2781. *Bookbinding.* Establishments primarily engaged in binding books printed in other establishments. Establishments primarily binding books printed in the same establishment are classified in Industry 273.

2782. *Blankbooks and Paper Ruling.* Establishments primarily engaged in paper ruling and in manufacturing blankbooks of all varieties, except loose-leaf.

2783. *Loose-leaf Binders and Devices.* Establishments primarily engaged in manufacturing loose-leaf binders and devices.

2789. *Miscellaneous Bookbinding Work.* Establishments primarily engaged in a wide variety of activities ancillary to bookbinding: bronzing, gilding, edging, deckling, embossing, and indexing, as well as others.

✓ ✓ ✓

279. *Service Industries for the Printing Trade.* Consists of four subindustries:

2791. *Typesetting.* Establishments primarily engaged in machine and hand typesetting for the trade, and in advertising typography.

2792. *Engraving and Plate Printing.* Establishments primarily engaged in engraving and etching steel and copper plates and in using these plates to print stationery, cards, invitations, maps, etc.

2793. *Photoengraving.* Establishments primarily engaged in preparing photoengraved plates. These establishments do not as a rule print from the plates which they make, but prepare them for others.

2794. *Electrotyping and Stereotyping.* Establishments primarily engaged in preparing electrotype and stereotype plates. These establishments do not generally print from the plates, but prepare them for others.

APPENDIX D

ADDITIONAL PRINTING AND PUBLISHING STATISTICS

Table D-1 Establishments and Employment in Printing and Publishing, New York Standard Metropolitan Area, 1954

Code number	Industry	Number of establishments	All employees	Production workers
2711	Newspapers	341	28,466	14,592
2721	Periodicals	579	26,771	4,211
2731	Books; publishing and printing	324	15,065	1,853
2732	Book printing	169	5,625	4,743
2741	Miscellaneous publishing	308	6,059	1,049
2751	Commercial printing	2,111	31,793	25,864
2761	Lithographing	538	13,685	10,637
2771	Greeting cards	108	5,353	4,005
2781	Bookbinding	256	6,858	6,235
2782	Blankbooks and paper ruling	97	1,000–2,499	[a]
2783	Loose-leaf binders and devices	40	2,019	1,656
2789	Miscellaneous bookbinding work	82	1,000–2,499	[a]
2791	Typesetting	283	4,700	3,789
2792	Engraving and plate printing	151	2,513	2,084
2793	Photoengraving	128	4,758	3,296
2794	Electrotyping and stereotyping	39	2,038	1,249

[a] Undisclosed.

Source: Special tabulation prepared by U.S. Bureau of the Census for the New York Metropolitan Region Study.

Table D-2 Printing Receipts for Selected Products in the Regions of the United States, 1954 (thousands of dollars)

Census region	Books and pamphlets	Periodicals	Catalogues, directories	Commercial printing and lithographing
United States, total	$304,894	$407,473	$125,264	$3,156,390
New England	33,985	29,704	7,553	189,866
Middle Atlantic, total	113,229	131,046	34,291	971,202
N.Y. State and N.J.	90,705	63,767	27,958	777,604
Pennsylvania	22,524	67,279	6,333	193,598
East North Central	88,941	169,597	61,892	1,053,444
West North Central	16,095	14,146	6,851	236,957
South Atlantic	19,305	20,338	3,347	205,950
East South Central	15,894	[a]	3,625	83,474
West South Central	4,226	[a]	2,635	100,636
Mountain	2,524	2,845	558	39,506
Pacific	10,376	14,596	4,512	275,355

Note: The percentage distribution of receipts by regions, based on these dollar figures, is found in Table 19 of the Printing and Publishing study.

[a] Less than one million dollars.

Sources and explanations

(1) Book and pamphlet printing and bookbinding: U.S. *1954 Census of Manufactures,* Vol. II, Part 1, Table 6B, p. 27A–14. This category includes the value of receipts for these activities whether done in establishments whose primary activity is book and pamphlet work or not; for example, some of this figure represents book printing done by establishments whose primary activity is lithographing.

(2) Magazines and periodicals, catalogues and directories: U.S. *1954 Census of Manufactures,* Vol. II, Part 1, Table 6C, p. 27B–19. These two product categories include only work done by commercial printing establishments, but the major part of these products are printed in such establishments: 90 per cent in the case of magazines and periodicals, and 65 per cent in the case of catalogues and directories.

(3) Commercial printing and lithographing: U.S. *1954 Census of Manufactures,* Vol. II, Part 1, Table 6B, p. 27B–17. These figures include the value of all products shipped by establishments whose primary activity was in these fields, *including* the products under (2) above, although *not* those under (1).

Table D-3 Actual and Projected Population, United States, Selected Years (in thousands)

	Total	Persons 15 and over
1929	121,770	85,565
1939	130,880	97,761
1947	143,446	105,579
1954	161,191	114,151
1965	194,102	139,275
1975	223,163	160,973
1985	249,942	180,266

Sources: For 1929–1947, Bureau of the Census, *Historical Statistics of the United States: 1789–1945* (Washington, 1949), p. 26. For 1954, Bureau of the Census, *Statistical Abstract of the United States: 1955* (Washington, 1955), p. 26. For projections, preliminary estimates developed by the New York Metropolitan Region Study. The final projections may differ slightly from these.

NOTES TO PRINTING AND PUBLISHING

CHAPTER I: A LOCATIONAL HISTORY

1. Lawrence C. Wroth, "Book Production and Distribution from the Beginning to the War Between the States," in *The Book in America,* ed. Hellmut Lehmann-Haupt (New York: R. R. Bowker, 1939), p. 101.

2. Frank Luther Mott, *A History of American Magazines* (Cambridge: Harvard University Press, 1939–1957), I, 200.

3. Earl L. Bradsher, "Book Publishers and Publishing," chapter XXIX in *The Cambridge History of American Literature,* IX (New York: Putnam's, 1921), 537–538.

4. Bliss Perry, *Park Street Papers* (Boston: Houghton Mifflin, 1908), p. 9. During its first fifteen years, roughly two-thirds of the *Atlantic's* contributors were from New England and much more than two-thirds of its pages were filled with their writing. Compare Mott, II, 495–496.

5. Mott, III, 5.

6. Theodore Peterson, *Magazines in the Twentieth Century* (Urbana, Ill.: University of Illinois Press, 1956), p. 3. This is an extremely useful source on the development of the modern magazine; we have relied heavily on it in the present account.

7. Mott, III, 5.

8. William S. Rossiter, "Printing and Publishing," *Twelfth Census of the United States, 1900,* Vol. IX, *Manufactures, Part III* (Washington, 1902), p. 1087. A perfecting press is one which prints on both sides of the paper simultaneously.

9. Theodore L. DeVinne, "American Printing," in *One Hundred Years of American Commerce,* ed. Chauncey M. Depew (New York: Haynes, 1895), pp. 318–319.

10. Cf. Charles W. Rantoul, Jr., "Paper and Pulp," *Twelfth Census of the United States, 1900,* IX, 1025, and L. H. Weeks, *A History of Paper Manufacturing in the United States, 1690–1916* (New York: Lockwood Trade Journal Co., 1916), p. 297.

11. The following magazines were among those moving to New York. After each title appears the previous location and date of the move: *Christian Examiner,* Boston, 1866; *North American Review,* Boston, 1878; *Cosmopolitan,* Rochester, N.Y., 1887; *Public Opinion,* Washington, D.C., 1895; *Arena,* Boston, 1899; *Woman's Home Companion,* Springfield, Ohio, 1901; *Lippincott's,* Philadelphia, 1914; *Dial,* Chicago, 1916; as well as a number of others whose names no longer have any meaning today.

12. Mott, IV, 83.

13. Hellmut Lehmann-Haupt, "Book Production and Distribution from 1860 to the Present Day," in *The Book in America* (see our note 1), p. 184.

14. Alfred Harcourt, "Publishing since 1900," in *Bowker Lectures on Book Publishing* (New York: R. R. Bowker, 1957), pp. 29–30. Some remarks in the same vein can be found in the memoirs of another publisher: George H. Doran, *Chronicles of Barabbas, 1884–1934* (New York: Harcourt, Brace, 1935), pp. 82–83.

15. Brander Matthews, "Our Monthly Gossip," *Lippincott's Magazine,* 37:106 (January 1886).

16. E. E. Pratt, *Industrial Causes of Congestion of Population in New York City* (New York: Columbia University Press, 1911), p. 76.

17. *American Trucking Trends* (Washington, D.C.: American Trucking Associations, Inc., 1956), p. 1.

CHAPTER 2: SERVING LOCAL MARKETS

1. *1954 Census of Manufactures,* Vol. II, Part 1 (Washington, 1957), Table 6C, p. 27A–16.

2. *1954 Census of Manufactures,* Vol. III, *Area Statistics* (Washington, 1957), p. 131–18.

3. In other cities listed, the average is misleading in the sense that there is usually one morning paper whose circulation outside the metropolitan area is head and shoulders above the rest of the papers in that city.

4. This situation was pointed out by A. F. Hinrichs, *The Printing Industry in New York and Its Environs* (New York: Regional Plan of New York and Its Environs, 1924), pp. 26–27.

5. The ratio of population in the central city to that of the Standard Metropolitan Area for cities we list in Table 7 are as follows:

New York City	66.8%
Chicago	65.9
Los Angeles	41.5
Philadelphia	56.4
Detroit	61.3
Boston	34.0
San Francisco	51.8

Data from Donald J. Bogue, *Population Growth in Standard Metropolitan Areas 1900–1950* (Washington, D.C.: Housing and Home Finance Agency, 1953), Appendix.

6. What is involved is a process of balancing gains in delivery time against the loss involved in printing in separate plants. This loss occurs because the economies of scale in newspaper printing continue well beyond the scale at which the largest papers are printed. Consequently, if part of the press run is printed at a separate plant, costs are necessarily higher than with a single plant. If the branch plant is large enough to achieve most of the economies

associated with large volume, however, the remaining differential in costs will be much smaller, and the savings in time and distribution cost will take over as the dominant factor.

7. This is seen by comparing union rates for identical jobs like hand compositors, machine operators, and mailers, city by city, in newspaper work and book and job printing. See U.S. Bureau of Labor Statistics, Bulletin No. 1194, *Union Wages and Hours, Printing Industry, July 1, 1955* (March 1956), Tables 9 and 13.

8. See the outspoken *Printing Is a Business,* by J. W. Rockefeller, Jr., and associates (Millburn, New Jersey: 1955), Ch. V.

9. See Appendix C for definitions of industries.

10. Strictly speaking, external economies, including the subcategory of concentration economies, are not restricted to the economies arising from the aggregate needs of the industry, but rather from the aggregate needs of the group of consumers using a given good or service. This group may of course cut across industry lines. For instance, companies which are organized to provide secretarial help on a short-term, short-notice basis are a concentration economy for firms in large cities in many industries at the same time. The external economies which we are particularly interested in, however, are those which are a function of the size of the industry; so we limit ourselves to that definition in the text.

11. Leonard A. Drake, *Trends in the New York Printing Industry* (New York: Columbia University Press, 1940), p. 119.

12. J. L. Stuart (Ampco Printing Co.), "A Case History in Plant Location," *Management in the Graphic Arts, 1954 Conference Proceedings* (Pittsburgh: Carnegie School of Printing Management, 1954), p. 46.

13. Committee on Price Determination of the Conference on Price Research, *Cost Behavior and Price Policy* (New York: National Bureau of Economic Research, 1943), Appendix D, "Specification of Book Paper Products," pp. 331–337. The designation "book paper" refers not to end use, but to its chemical composition.

14. Figure from Sal Iaccio of the International Association of Machinists.

Chapter 3: Periodicals and Books

1. Nathan Belfer, on assignment from the New York Metropolitan Region Study, prepared a memorandum on the periodicals industry which the author drew upon in writing this section of the chapter.

2. Computed from *1947 Census of Manufactures* (Washington, 1949), Vol. II, *Statistics by Industry,* p. 348, and Vol. III, *Statistics by States,* p. 429.

3. Computed from *1954 Census of Manufactures* Vol. II, Part 1 (Washington, 1957), p. 27A–4, and Vol. III, *Area Statistics,* p. 131–18.

4. These are magazines that are members of the Audit Bureau of Circulation. Very few large magazines are not.

5. See *Special Library Resources,* ed. Isabel L. Townes (New York: Special

Libraries Association, 1947) for a listing of the amazing variety with accompanying reference personnel.

6. Estimate by Lawrence Ladler, president of the Society of Magazine Writers.

7. See "One-Shot Magazines Spread Fast, Explore Hi-Fi, Flowers, Elvis," *Wall Street Journal*, Nov. 6, 1957.

8. "Externally conditioned labor" is labor which is available at a given site to some degree independently of the inducements offered. New England's textile mill towns used the female labor which was tied to the place because of the jobs of their husbands. University towns usually have high-grade, low-cost secretarial help because of student wives who are not free to seek higher-paying work in the Big City.

9. In addition to the ownership of special facilities and so on, there is also the fact that an "irrational" location can be perpetuated almost indefinitely by a firm with some monopoly power or a differentiated product.

10. Figure for 1954 from the *1954 Census of Manufactures*, Vol. II, Part 1 (Washington, 1957), p. 27A–11, Table 6A.

11. Cf. A. F. Hinrichs, *The Printing Industry in New York and Its Environs* (New York: Regional Plan of New York and Its Environs, 1924).

12. Leonard A. Drake, *Trends in the New York Printing Industry* (New York: Columbia University Press, 1940), Ch. IV. The twelve largest were determined by ascertaining the peak employment of each firm during the period.

13. Computed from *1954 Census of Manufactures*, Vol. II, Part 1 (Washington, 1957), pp. 27B–6 and 27B–19.

14. Drake, p. 98. A rotary press is one which prints from curved plates, as opposed to the considerably slower flat-bed press. In a web press the paper is fed from a roll, rather than from a stack one sheet at a time.

15. The reason for having multiple printing plants is, in the case of *Time*, fundamentally the need for fast dissemination of the news. *Time* also saves money on postage by using printers in different parts of the country, but this cost reduction is more than offset by the additional manufacturing costs. If cost savings alone were the objective, it would be desirable to print all copies on a single press.

16. Source: Table 1, p. 138.

17. There is, strangely, relatively little written about the book publishing industry, in spite of the fact that the industry itself is concerned with the dissemination of information. From the point of view of the location of the industry, there is nothing available but a few scattered comments. Certain books —all of them rather chatty—have been useful for background information and factual material, however. They are: William Miller, *The Book Industry* (New York: Columbia University Press, 1949); *The Book in America*, ed. Hellmut Lehmann-Haupt (New York: R. R. Bowker, 1939); *What Happens in Book Publishing*, ed. Chandler B. Grannis (New York: Columbia University Press, 1957); and *Bowker Lectures on Book Publishing* (New York: R. R. Bowker,

1957). In addition, *Literary Market Place* (New York: R. R. Bowker, 1957) provided much information not available elsewhere about various aspects of the publishing industry.

18. This separation of figures was not repeated until 1939, when it became a regular feature of Census reports.

19. See Miller (note 17 above), p. 6.

20. *1954 Census of Manufactures,* Vol. II, Part 1 (Washington, 1957), p. 27A–23.

21. Lehmann-Haupt, "Book Production and Distribution from 1860 to the Present Day," in Lehmann-Haupt (note 17 above), p. 185.

22. See *Literary Market Place* (note 17 above), pp. 79–93.

23. This was true at a much earlier stage in the industry as well. "Probably a majority of houses did not at any time after the Civil War do their own printing. Approximately 18 per cent of those listed in the Census of 1914 had plants, and these accounted for 31 per cent of the total production." Donald Sheehan, *This Was Publishing* (Bloomington, Ind.: Indiana University Press, 1952), p. 21. Unfortunately, up-to-date statistics on the point are not available.

24. Computed from Appendix D, Table D–2.

CHAPTER 4: PRINTING COSTS

1. For details of the system, see U.S. Federal Trade Commission, Docket No. 3760, "Findings of Fact and Conclusions" (Washington, June 30, 1945, mimeo.), pp. 13–14. The FTC issued a Cease and Desist Order against the system in 1945, but it has not been effective. The FTC prohibited *collusion* on maintaining zoned prices, but did not enjoin the manufacturers individually from continuing the old system by themselves with the same result. This particular defect has been remedied in another similar case involving manufacturers of chains, by prohibiting individual firms from continuing zoned delivered pricing. See *Chain Institute,* Federal Trade Commission Docket No. 4878.

2. The Census category of cost of materials includes the cost of contract work—that is, typesetting, photoengraving, and the like. This work amounts to a bit more than one-eighth of the total materials figure. We have left the cost of contract work in the materials figure because it is not important from our point of view. Costs for contract work will vary in much the same way as do the costs for printing proper, since the principal cost component is labor, to an even greater extent than in printing proper.

3. George A. Stevens, *New York Typographical Union No. 6, Study of a Modern Trade Union and Its Predecessors,* Annual Report of the Bureau of Labor Statistics, New York State Department of Labor, 1911, pp. 34–35.

4. Stevens, p. 64.

5. Elizabeth Faulkner Baker, *Printers and Technology* (New York: Columbia University Press, 1957), p. 259.

6. See Jacob Loft, *The Printing Trades* (New York: Farrar & Rinehart,

1944), pp. 162–163, and National Recovery Administration, *Codes of Fair Competition*, Vol. VII, pp. 29–31.

7. U.S. Bureau of Labor Statistics, Bulletin No. 1194, *Union Wages and Hours, Printing Industry, July 1, 1955* (March 1956), pp. 33–34.

8. The concept of the key rate is discussed by John T. Dunlop, "The Task of Contemporary Wage Theory," in *New Concepts in Wage Determination*, ed. George W. Taylor and Frank C. Pierson (New York: McGraw-Hill, 1957), pp. 129–130: "A job cluster is defined as a stable group of job classifications of work assignments within a firm . . . which are so linked together by (1) technology, (2) the administrative organization of the production process, . . . or (3) social custom that they have common wage-making characteristics. . . . The wage rates for the operations and jobs within a cluster are more closely related in their wage movements and wage-making forces than are rates outside the cluster. . . . Ordinarily a job cluster will contain a key rate, or in some cases several. The cluster consists of the key rate(s) and a group of associated rates. The key rate may be the highest paid, or the rate paid at the top step in a promotion ladder, or the rate paid for a job at which a large number of workers are employed. Typically, the key-rate jobs show relatively less change in job-content over a period of time and are often relatively more standardized among firms than are other jobs. The key rates are those which managements and unions typically have in mind and explicitly discuss in considering the internal wage structure."

9. Emily Clark Brown, "Book and Job Printing," in *How Collective Bargaining Works* (New York: Twentieth Century Fund, 1942), p. 156.

10. Loft, p. 286.

11. "Extent and Nature of Collective Bargaining," *Monthly Labor Review*, 64:765–769 (May 1947).

12. Sumner H. Slichter, *Union Policies and Industrial Management* (Washington, D.C.: Brookings Institution, 1941), pp. 192–193.

13. Brown, p. 161.

14. The history of rates for all classes of mail is available in *United States Domestic Postage Rates, 1789–1956*, Post Office Department Publication 15 (Washington, 1957). Interpretative material on the rate structures is available in the following two sources, both of which have been very useful in the present paper: Jane Kennedy, "Structure and Policy in Postal Rates," *Journal of Political Economy*, 65:185–208 (June 1957); and *Report on Second-Class Mail to the Postmaster General*, submitted by Charles A. Heiss (Washington, 1946).

15. Kennedy, p. 189. For many of the large national periodicals, the percentage was frequently much higher.

16. Cf. the articles on "The History of Magazine Distribution," by Roy Quinlan, in *Magazine World*, March–May 1946.

17. *Postal Rate Revision*, Hearings before the Committee on Post Office and Civil Service, House of Representatives, 80th Cong., 1st sess., on

H.R. 2408 (Washington, 1947), p. 400; and Lowell Shumway, "Circulation Methods," *Magazine Industry,* 1:21–24 (Winter 1950).

18. See Chauncey D. Harris, "The Market as a Factor in the Location of Industry in the United States," *The Appraisal Journal,* 24:67 (January 1956). This paper was reprinted from the *Annals of the Association of American Geographers,* December 1954.

19. Rather than dealing with figures on the geographical distribution of printing by industry, it makes more sense to look at product statistics, although they are not available on as fine an area basis as the industry figures. The difference is the following: industry figures (which we have used so far) classify production by broad categories. In particular, "commercial printing" covers a multitude of varied activities. But if we break down the figures by product categories, we can see the locations of the types of printed products which are most sensitive to transport costs. The two geographical areas of particular importance are the Middle Atlantic States and the East North Central States (Ohio, Indiana, Illinois, Michigan, and Wisconsin). In printing, East North Central means largely Ohio and Illinois, in particular the Chicago metropolitan area.

20. Report of the Interstate Commerce Commission, printed in *Adjustment of Postal Rates,* Hearings before the Committee on Post Office and Civil Service, Senate, 82d Cong., 1st sess., on S. 1046, S. 1335, and S. 1369 (Washington, 1951), p. 908.

21. Figures calculated from U.S. *1954 Census of Manufactures,* Vol. II, Part 1 (Washington, 1957), pp. 27B–6 and 27B–19.

22. These figures do not include advertising pamphlets and circulars. A pamphlet, for Census purposes, has more than 8 but less than 64 paperbound pages. The figures also do not include book printing and binding done in establishments which are classified in the book publishing industry. This fraction is relatively small, although exact figures are not available.

23. *Adjustment of Postal Rates* (note 20, above), p. 912.

24. *United States Domestic Postal Rates* (note 14, above), p. 10.

25. Statement by J. R. Brackett, general manager of Printing Industry of America, in *Postal Rate Revision* (note 17, above), p. 444.

26. *Ibid.*

CHAPTER 5: THE NEXT TWENTY-FIVE YEARS

1. Raymond B. Nixon, "Who Will Own the Press in 1975?" *Journalism Quarterly,* 32:12 (Winter 1955).

2. See *Printers' Ink,* October 31, 1958, Section 2, p. 201.

3. Examples of these developments are an inexpensive engraving process, enabling small papers to afford news pictures, and new photocomposition methods, permitting inexpensive typesetting with a wider variety of type faces. Developments in presses which will substantially reduce cost for news-

papers with relatively short runs are also in the offing. See Mel Most, "Tomorrow's Newspaper: Technological Progress," *Barron's*, December 15, 1958, pp. 3, 15–19.

4. Nixon, p. 13.

5. 1945 figure computed from Bureau of the Census, *Historical Statistics of the United States: 1789–1945* (Washington, 1949), p. 26. 1956 figure from Bureau of the Census, *Statistical Abstract of the United States: 1957* (Washington, 1957), p. 24.

6. These percentages are computed from the population projections presented in Appendix D, Table D–3. Other projections are made by the Bureau of the Census on a variety of different fertility assumptions. None of the Census estimates shows a substantial rise over the 1954 percentages. The only one which shows *any* rise is the prediction under the "high fertility" assumption, which the Bureau labels as questionable. Under the "high fertility" assumption, the under-15 group is expected to rise to 30.9 per cent by 1965 and to 31.5 per cent by 1975. These figures seem to be an upper limit. The projections are presented in the *Statistical Abstract* for 1957, Table 3.

7. Figures computed from U.S. Dept. of Commerce, *Business Statistics* (Washington, 1957), p. 43, and *Survey of Current Business*, 38:S8–S9 (March 1958).

8. See *Printers' Ink*, Oct. 31, 1958, Section 2, p. 204.

9. Much of this technological development is discussed in the article by Most, referred to in note 3, above.

10. Leo Bogart reports these surveys in "Magazines since the Rise of Television," *Journalism Quarterly*, 33:153–166 (Spring 1956), especially pp. 154–155.

11. See *Printers' Ink*, Oct. 31, 1958, Section 2, p. 233.

12. Advertising linage figures are presented in U.S. Dept. of Commerce, *Business Statistics* (Washington, 1957). Recent statistics are in the *Survey of Current Business*.

13. The content analysis is reported in Bogart, p. 159.

14. The Regional Plan Association estimates that the Region's population will be 9.1 per cent of the nation's in 1965, and 8.9 per cent in 1975. A straight-line extrapolation of this trend gives us a figure of 8.6 per cent for 1985. See *People, Jobs, and Land, 1955–1975*, RPA Bulletin Number 87 (June 1957).

15. Data available for recent years, collected by the state departments of labor, suffer from certain deficiencies of coverage and classification which prevent our using them in this particular context.

Electronics

APPENDIX E

ELECTRONICS PROJECTIONS—SOURCES AND METHODS

This appendix describes the methods by which estimates of future levels of output and employment in the electronics industry were derived. To begin with, we adopted the projections of the staff of the New York Metropolitan Region Study with respect to future national levels of population and Gross National Product. These estimates, the basis for which will be described in the final volume of this series, are shown in Table E–1. Beyond that, we relied to a considerable extent upon demand characteristics used by Sylvania Electric Products, Inc., in its projections for the year 1965.* The Sylvania staff is continually revising its estimates; the figures we used were the latest available at the time our electronics study was prepared.

Table E–1 Projected Levels of Population and Gross National Product, United States

	1965	1975	1985
Population (millions)	194	223	250
Gross National Product [a] (billions of dollars)	600	879	1,293

[a] GNP is in 1957 dollars and is based on an assumption of a 2½ per cent annual increase in productivity.

Our estimates therefore represent for the most part a wedding of New York Metropolitan Region Study projections of national aggregates and Sylvania projections of demand characteristics.

* The Sylvania forecasts are used with the permission of Sylvania Electric Products, Inc. They were prepared under the direction of Frank W. Mansfield, Director of Marketing Research.

✓ Consumer electronics

Home radio and television. The demand in a given year for a durable good like a radio or television set consists of two parts: replacement demand and initial demand. When radio was first invented and later when television appeared, the demand in a given year was largely made up of initial demand. Later, replacement demand grew in importance. This latter type of demand, while sensitive to such factors as income and technological obsolescence, is largely a function of history. It depends in part on the stock of sets which has been built up over the years and the age of that stock. Thus in projecting replacement demand in 1965 we need to project how large the stock of sets will be at the beginning of that year and how much of that stock will be replaced during the year.

But there will also be initial demand, because of (1) population growth and (2) an increase in the number of sets per household. The latter, in turn, depends upon the growth of income and such developments as portable television, color television, transistor radios, and so on.

The critical estimates for 1965 which we adopted from the work of Sylvania are the following: that replacement demand in that year will be 7.2 per cent of the existing stock for radio and 8.1 per cent for television; that 98.6 per cent of households will have at least one radio and 97.1 per cent at least one television set; that the average number of radios per radio home will be 2.00 and the average number of television sets per television home will be 1.289.

On the basis of New York Metropolitan Region Study projections for population growth, the number of households at the end of 1964 will be 57,900,000, and 900,000 more will be added during 1965. The number of radio sets in existence at the beginning of 1965 will be 113,377,000. At a 7.2 per cent rate of replacement, replacement demand in 1965 will equal 8,163,000 sets. The demand of 900,000 additional households along with the increased demand on the part of existing households will require an additional 2,577,000 radios. The total demand will therefore equal 10,740,000 sets. Similar calculations for television yield a total demand for 8,555,000 television sets. The relevant figures are shown in Table E–2.

Automobile radio. The demand for automobile radios is based on the demand for automobiles and the ratio of automobiles with radios to total automobiles. Sylvania estimates that in 1965 sales of domestically produced automobiles will be 7,330,000 and that 6,450,000 radios will be sold to the public in this form.

Table E-2 Projections of Radio and Television Demand (home sets)

National aggregates	
1. Households, beginning of 1965	57,900,000
2. New households, 1965	900,000
3. Households, end of 1965	58,800,000
Radio demand	
1. Ratio of radio households to all households, beginning of 1965	98.4%
2. Ratio of radio households to all households, end of 1965	98.6%
3. Radio sets per radio household, beginning of 1965	1.99
4. Radio sets per radio household, end of 1965	2.00
5. Replacement rate, 1965	7.2%
6. Stock of radio sets, beginning of 1965	113,377,000
7. Replacement demand, 1965	8,163,000
8. Initial demand, 1965	2,577,000
9. Total demand, 1965	10,740,000
Television demand	
1. Ratio of TV households to all households, beginning of 1965	96.4%
2. Ratio of TV households to all households, end of 1965	97.1%
3. TV sets per TV household, beginning of 1965	1.268
4. TV sets per TV household, end of 1965	1.289
5. Replacement rate, 1965	8.1%
6. Stock of TV sets, beginning of 1965	70,774,000
7. Replacement demand, 1965	5,734,000
8. Initial demand, 1965	2,821,000
9. Total demand, 1965	8,555,000

Other consumer products. Sylvania estimates phonograph sales (including hi-fi) for 1965 at 5,200,000 sets. This figure does not include phonographs in combination with TV or radio sets, which are already included in our radio and TV figures. We have adopted this estimate. Similarly we have adopted Sylvania's estimate of the demand for other items such as tape recorders, magnetic tape, and electronic organs. Sylvania also projects a large increase in sales of phonograph records, but since we have not included records as a part of the electronics industry, we exclude them from our projections.

Aggregate expenditure on consumer products. Our next step is to translate the above projections of volume into values at 1956 prices. Values, as we shall see, are the only common denominator for describing the level of activity of the various electronics branches. In television we would be misled by today's average price because it is expected that a larger percentage of sets will be of the color variety in the future. According to Sylvania's estimates, the proportions will be as shown in Table E-3. If the present price structure were to be maintained, the

Table E-3 Projections of Proportions of Total Television Sales in Various Categories

	1956	1965
Television sales, total	100%	100%
Black and white sets, total	98	82
Color sets	2	18

average price of TV would rise, since color now sells at more than twice the price of black and white. On the other hand, as color sales increase, their price will fall relative to black and white because of economies of mass production. This aspect of productivity must be taken into account at this stage, though other elements in the adjustment for productivity changes are included implicitly in later stages of our calculations.

Sylvania projects a price relationship of 2 : 1 in 1965 for color sets versus black and white sets. Black and white sets they expect will be valued at $111 while color sets will be valued at $230 per set. This means that if the price of color sets in 1956 had already reflected the productivity gains anticipated by 1965, the price of these color sets would be roughly $250. Therefore, the value of television production in 1965 can be computed as 82 per cent at $111 per set and 18 per cent at $230 per set. This means that sales of all TV sets, projected at 8,555,000 in 1965, will have a value of $1,133,000,000.

We provide no similar adjustments in radio and phonograph sales, despite the fact that there will surely be changes in the mix as between different types of radios and phonographs. The data are not available to permit such adjustments but we can derive some comfort from the fact that the variations in price are not great and the changes in demand will not be as marked as in television. For radio, the average value of $21.50 per set multiplied by the projected 1965 total number of 10,740,000 yields a total value of $231,000,000. Auto radio adds $161,000,000, and phonographs—at $38 per set—an additional $198,000,000. These figures along with those for other types of consumer products are brought together in Table E-4.

Table E-4 Projected Expenditures for Consumer Electronics in 1965 at "1956 Prices" (millions of dollars)

Television	1,133
Home radios	231
Auto radios	161
Phonographs	198
Electronic organs	25
Magnetic tape and tape records	62
Tape recorders	80
Hi-fi components	50
Total	1,940

For the years 1975 and 1985, the figures given in the final chapter of our case study on the electronics industry were derived by a linear extrapolation of the trend in the absolute level of expenditures as manifested in recent years and as projected for 1965. However, by taking account of the projected growth in population and assuming 100 per cent saturation for both radio and television, with something like 2 radio sets and 1.5 TV sets per household, one would arrive at figures roughly approximating the projections of sales given in Table 11 of the electronics study.

MILITARY ELECTRONICS

Our basic assumptions are the following: that defense expenditures will be equal to 9 per cent of Gross National Product and that electronics will take 14 per cent of defense expenditures. These assumptions, when applied to the GNP projections given in Table E-1, yield the figures shown in Table E-5.

Table E-5 Projections of Expenditures on Military Electronics (billions of dollars)

	1965	1975	1985
Defense expenditures	54.0	79.0	116.0
Military electronics expenditures	7.6	11.1	16.2

INDUSTRIAL ELECTRONICS

Sylvania has estimated the prospective use of digital computers on the basis of the following assumptions. Only firms with more than 100 employees need to be considered as potential customers. Only those with more than 500 employees are potential customers for large or medium-size computers. And only those with assets of more than $200,000,000 can afford large-size computers. These assumptions lead to the projections shown in Table E-6. It should be noted that such projections make no allowance for the possibility that new customers may be created through growth or merger.

Table E-6 Prospective Market for Initial Installation of Digital Computers

Type of computer	Number of firms	Number of computers	Value of computers (millions of 1956 dollars)
Large	305	995	1,790
Medium	7,795	7,795	1,170
Small	43,500	43,500	1,088
Totals	51,600	52,290	4,048

In addition to the $4 billion shown in Table E-6, Sylvania estimates an expenditure of $2 billion for companion equipment. Thus the total prospective use is estimated at $6 billion. In Sylvania's view this demand will have been satisfied by 1968, so that we can only look to replacements to generate current demand in the years following. But because of the rapid rate of innovation in this field Sylvania assumes a high annual rate of replacement, 15 per cent, giving rise to expenditures of $900,000,000 a year.

We are inclined to assume that the initial demand will not be satisfied as early as 1968. The introduction of digital computers into the operations of a firm requires some fundamental adjustments which take a good deal of time. Habits of thinking must be changed: executives have to learn to appreciate what these machines are capable of doing for them, and must reshuffle their staff structures to make most efficient use of both personnel and equipment. It is not likely, therefore, that the rate of saturation will be approached with anything like the speed we have experienced in consumer items, such as television.

In projecting the demand for industrial electronics in 1965, therefore, we have assumed that the process of filling the initial demand for digital computers will be going on for a long time. Our projection for 1965 sales of computers—both initial and replacement—is $1.5 billion. We expect that other types of industrial electronics will provide a market of about $2.0 billion. Such items as mobile radio, microwave, medical equipment, scientific instruments, automation, and nuclear control are developing quickly. For 1975 and 1985 we have projected a further rapid growth of demand, though not at the same rate.

✓ Replacement components

Sylvania projects total sales of replacement components at $1.1 billion for 1965. We have adopted this figure and have projected a somewhat more rapid increase for 1975 and 1985 on the grounds that there will have been a substantial growth in the stock of electronic equipment.

✓ Productivity and employment

It is obviously impossible, in view of the changing mix of products, to find a way of relating the physical output of the electronics industries to their employment. What we have tried to do therefore is to answer this question: What will be the relationship of sales to employment in future years?

By using annual figures for employment and payrolls in the nation as a whole we found fairly persistent relationships which provide some support for two critical assumptions which we are forced to make: first, that payrolls in electronics will rise roughly at the same rate as sales of electronic products; and second, the earnings per worker in electronics will rise roughly at the same rate as earnings per worker in all manufacturing. Table E–7 shows that payrolls in electronics have averaged around 40 per cent of sales. Table E–8 shows that earnings per worker in elec-

Table E-7 Payrolls in Electronics Industries as Percentage of Sales of Electronic Products

1947	42	1952	39
1948	n.a.	1953	41
1949	n.a.	1954	37
1950	39	1955	37
1951	38	1956	39

Source: Payrolls through 1954 from U.S. *1954 Census of Manufactures;* 1955 and 1956 payrolls from U.S. *Annual Survey of Manufactures* for those years. The figures cover four industries, Standard Industrial Classifications 3661, 3662, 3571, 3811. Electronics sales from *1958 Electronics Industry Fact Book* (Electronic Industries Association, Washington, D.C., 1958).

Table E-8 Index of Average Annual Earnings in Electronics Industries and All Manufacturing, 1947–1956 (1950 = 100)

	Electronics	All manufacturing
1947	85	86
1948	n.a.	n.a.
1949	n.a.	n.a.
1950	100	100
1951	112	111
1952	120	117
1953	120	124
1954	126	126
1955	131	130
1956	140	138

Source: Figures through 1954 from U.S. *1954 Census of Manufactures;* figures for 1955 and 1956 from U.S. *Annual Survey of Manufactures* for those years. The figures cover four industries, Standard Industrial Classifications 3661, 3662, 3571, 3811.

tronics industries have risen at roughly the same rate as earnings in all manufacturing industries.

All that remains is to project the rate at which earnings per worker will rise. Our Gross National Product projections were based on the assumption of a 2½ per cent annual rise in productivity. This is the figure we shall adopt here. Therefore, average earnings in electronics industries are expected to rise by 25 per cent between 1956 and 1965, by 60 per cent between 1956 and 1975, and by 105 per cent between 1956 and 1985. In Table E–9 we show how we arrived at the final projections of employment.

Table E–9 Projected Sales, Payrolls, Payroll per Worker, and Employment in Electronics Industry

	1965	1975	1985
Total sales (in billions of dollars)	14.0	20.8	29.8
Total payrolls (in billions of dollars)	5.6	8.3	11.9
Payroll per worker (in dollars)	5,604	7,173	9,190
Employment	1,000,000	1,150,000	1,295,000

NOTES TO ELECTRONICS

CHAPTER 1: INTRODUCING AN INDUSTRY

1. O. E. Dunlap, Jr., *Radio's 100 Men of Science* (New York: Harper & Bros., 1944), p. 167.

2. J. Gordon Cook, *Electrons Go to Work* (New York: Dial Press, 1957), p. 171. This whole book is useful to a layman seeking to grasp the meaning of electronics.

3. See Table 1, p. 247.

4. See Table 1, including note a.

5. J. T. Cooper, *Thomas A. Edison* (New York: Frederick A. Stokes, 1941), especially pp. 71, 84.

6. J. W. Hammond, *Men and Volts* (New York: Lippincott, 1941), pp. 42–43, 113 ff., 156, 173. Other sources on Edison's locations: H. C. Passer, *The Electrical Manufacturers, 1875–1900* (Cambridge: Harvard University Press, 1953), pp. 58, 85; Cooper, p. 207; G. S. Bryan, *Edison, the Man and His Work* (New York: Knopf, 1926), p. 61.

7. Cook (see note 2 above), pp. 89–90. On Thomson, see Cook, pp. 18, 28–29.

8. W. R. Maclaurin, *Invention and Innovation in the Radio Industry* (New York: Macmillan, 1949), pp. 45–47; Cook, 90–94.

9. Dunlap (note 1, above), p. 168. Other sources on de Forest include Maclaurin, p. 80; Cook, pp. 94–96.

10. Maclaurin, p. 60.

11. On Steinmetz, see Hammond (note 9, above), p. 199. On Langmuir, see Dunlap, p. 202.

12. H. L. Jome, *Economics of the Radio Industry* (New York: A. W. Shaw Co., 1925), p. 11; E. E. Bucher, "A Resumé of Early Radio Development," in *The Radio Industry* (New York: A. W. Shaw Co., 1928), p. 59.

13. Hammond (note 9, above), pp. 352–354; Jome, p. 47; Maclaurin, p. 94.

14. Jome, p. 47.

15. The first broadcasting station in the New York area was WJZ, owned jointly by RCA and Westinghouse. It started operations in Newark in 1921 and soon moved to midtown Manhattan to be nearer the performers. A. T. & T. opened its own station, WEAF, in Manhattan's financial district. See G. L. Archer, *History of Radio to 1926* (New York: American Historical Society, 1938), pp. 293–305. Broadcasting was slow in becoming a commercial success, though by 1925 there were 500 licensed broadcasting stations in the United States. A. T. & T., which had spent more than any other firm on radio development, held patents giving it the commercial rights to radio-telephony. In 1926 RCA paid one million dollars to the telephone company

for WEAF and an agreement that the telephone company would leave the broadcasting field.

16. *1958 Electronics Industry Fact Book* (Electronic Industries Association, Washington, D.C., 1958), p. 9.

17. For a full account see Maclaurin (note 8, above), Ch. IX.

18. *Ibid.*, p. 255. The Allen B. DuMont Laboratories in Clifton, New Jersey, manufactured and sold television sets to the public just prior to World War II, but the war soon intervened. *Ibid.*, p. 219.

19. *1958 Electronics Industry Fact Book*, p. 14.

20. *Ibid.*, pp. 9 and 4.

21. *Ibid.*, p. 28.

Chapter 2: electronics in the living room

1. *Radio Broadcast*, June 1929, and *Moody's Industrials, passim.*

2. Paul Schubert, *The Electric Word, The Rise of Radio* (New York: Macmillan, 1928), p. 239.

3. J. L. Ray, "The Distribution and Merchandising of Radio Equipment," in Harvard University Graduate School of Business Administration, *The Radio Industry: The Story of Its Development* (New York: A. W. Shaw Co., 1928), p. 246.

4. H. L. Jome, *Economics of the Radio Industry* (New York: A. W. Shaw Co., 1925), p. 73.

5. See Chart 2 in this chapter. Strictly speaking, it is estimated that 400 firms were in the industry during 1925, though not on any one date.

6. Schubert, p. 243. For a time an independent radio group was active, manufacturing under patents awarded to Professor L. A. Hazeltine of Stevens Institute of Technology, Hoboken, New Jersey. Radio Frequency Laboratories in Boonton, New Jersey, was also offering licenses on its independently developed patents.

7. H. Gernsback, "Edison and Radio," *Radio News*, December 1926, p. 625.

8. *Moody's Industrials*, 1925–1929, *passim.*

9. The average retail price of radio sets went from a high of $124 in 1927 to $82 in 1930 and to $14 in 1933. The last figure reflected the importance of the midget set. *Electronics*, March 1931, p. 535, and July 1933, p. 197.

10. *Electronics*, April 1931, p. 578.

11. *Electronics*, March 1931, p. 533.

12. W. R. Maclaurin has estimated market shares in 1940 for radio in which Philco, for example, is almost tied with RCA and five times as great as GE. See his *Invention and Innovation in the Radio Industry* (New York: Macmillan, 1949), Table X, p. 146.

13. A wider range can also be quoted, based on reports from 16 plants. "Labor costs attributable to the assembly of a set are usually less than 20 per cent of total costs and often as low as four per cent." *Radio and Television Manufacturing* (Bureau of Labor Statistics, Washington, D.C., February

1952), p. 2. The high figure undoubtedly comes from a highly integrated plant or a maker of custom sets.

14. A breakdown of skill requirements for radio manufacturing appears in Bureau of Labor Statistics, *Wage Structure, Radios, 1945,* Series 2, No. 19, Table 3. Also see *Defense Manpower Requirements* (Bureau of Labor Statistics, Manpower Report No. 12, February 1952), p. 35; *Employment Outlook in Electronics Manufacturing* (Bureau of Labor Statistics, Bull. 1072, February 1952), pp. 19 and 21; and *Radio and Television Manufacturing* (note 13, above), p. 1.

15. That trucking is used extensively in the distribution of radio and television sets is also revealed by the Transportation Survey referred to in Table 5, p. 273.

16. *1958 Electronics Industry Fact Book* (Electronic Industries Association, Washington, D.C., 1958), pp. 12–13.

CHAPTER 3: THE BITS AND PIECES

1. *1958 Electronics Industry Fact Book* (Electronic Industries Association, Washington, D.C., 1958), p. 22.

2. *Electronic Industries and Tele-Tech,* June 1957, pp. 368–369.

3. *1958 Electronics Industry Fact Book,* pp. 21, 22, 23. The percentage is computed from the following dollar figures appearing on those pages: all electron tubes $970 million; receiving tubes $384 million; picture tubes $183 million.

4. From U.S. *1947 Census of Manufactures* and special 1954 tabulation made by the Census Bureau for the New York Metropolitan Region Study.

5. *Hearings,* "Learners in the Small Electrical Products Industry," Wage and Hour and Public Contracts Divisions, U.S. Department of Labor, 1956, *passim.*

6. Later, as the tube market contracted, this 320,000-square-foot plant was used first as a warehouse and then sold. *Wall Street Journal,* January 24, 1958.

7. Based on 1957 unit production, by end-use, in *1958 Electronics Industry Fact Book,* p. 21.

8. One executive reported, however, that he was very pleasantly surprised at the availability and quality of tool-and-die work in the South.

CHAPTER 4: MILITARY AND INDUSTRIAL ELECTRONICS

1. George O. Von Frank, *The Electronics Industry in New York State* (Department of Commerce, State of New York, Albany, 1957), p. 14.

2. Von Frank, pp. 12–14.

3. *1958 Electronics Industry Fact Book* (Electronic Industries Association, Washington, D.C., 1958), p. 28.

4. Von Frank, pp. 26–28.

5. *Defense Manpower Requirements in Electronics Production,* Manpower Report No. 12 (Washington: Bureau of Labor Statistics, 1952), p. 27.

6. U.S. *1954 Census of Manufacturers,* Vol. II, Part 1, p. 38A–13.

7. *Defense Manpower Requirements,* p. 27. Precise figures are not available for more recent years, but there is no question that large metropolitan areas still have the bulk of the employment.

8. For New York City's concentration, see Table 10, next chapter. As for size of plants in this industry, the average number of employees per establishment in the nation in 1954 was 122 (see U.S. *1954 Census of Manufactures,* Vol. II, Part 1, p. 38A–4). The average in the New York Metropolitan Region was 228, according to an unpublished tabulation by the U.S. Census Bureau.

9. Not all of the 3,500 are active accounts at any given time. Another convenient example of testing services can be found in Florida, where electronics manufacturing has had a rapid growth, and where there has recently been established what is called "the most complete commercial laboratory for servicing and calibrating electronics and nucleonics instruments in the southeastern part of the country." *Electronic News,* March 27, 1958.

10. *Defense Manpower Requirements,* p. 25.

11. *Electronics,* June 13, 1958, p. 14.

12. *Electronic News,* Oct. 7, 1957. See also note 6, above, relating to a testing laboratory in Florida.

13. E. T. VanDeusen, "Mr. Fogg's Remarkable Electronics Factory," *Fortune,* January 1956, pp. 113 and 115.

14. *Electronic News,* March 31, 1958.

CHAPTER 5: THE SHAPE OF THINGS

1. U.S. *1954 Census of Manufactures,* including unpublished special tabulations.

2. *1958 Electronics Industry Fact Book* (Electronic Industries Association, Washington, D.C., 1958), p. 15.

3. Cf. *Electronic News,* Dec. 30, 1957.

4. Nat Hentoff, "The FM Boom: Radio for Grown-Ups," *The Reporter,* May 1, 1958, p. 34.

5. Reported in *Electronic News,* March 17, 1958.

6. Frank W. Mansfield, Director of Marketing Research, Sylvania Electric Products, Inc., speech before New York Society of Security Analysts, Nov. 8, 1956 (mimeo.), p. 7.

Index

N
W
E
S
NEW
NEW YORK
DUTCHESS
ORANGE
PUTNAM
FAIRFIE
WESTCHESTER
ROCKLAND
Delaware River
Hudson River
PASSAIC
NEW JERSEY
BERGEN
Long
MORRIS
BRONX
ESSEX
NEW YORK
HUDSON
QUEENS
NASSAU
UNION
KINGS
RICHMOND
SOMERSET
MIDDLESEX
MONMOUTH
Atlantic
Delaware River

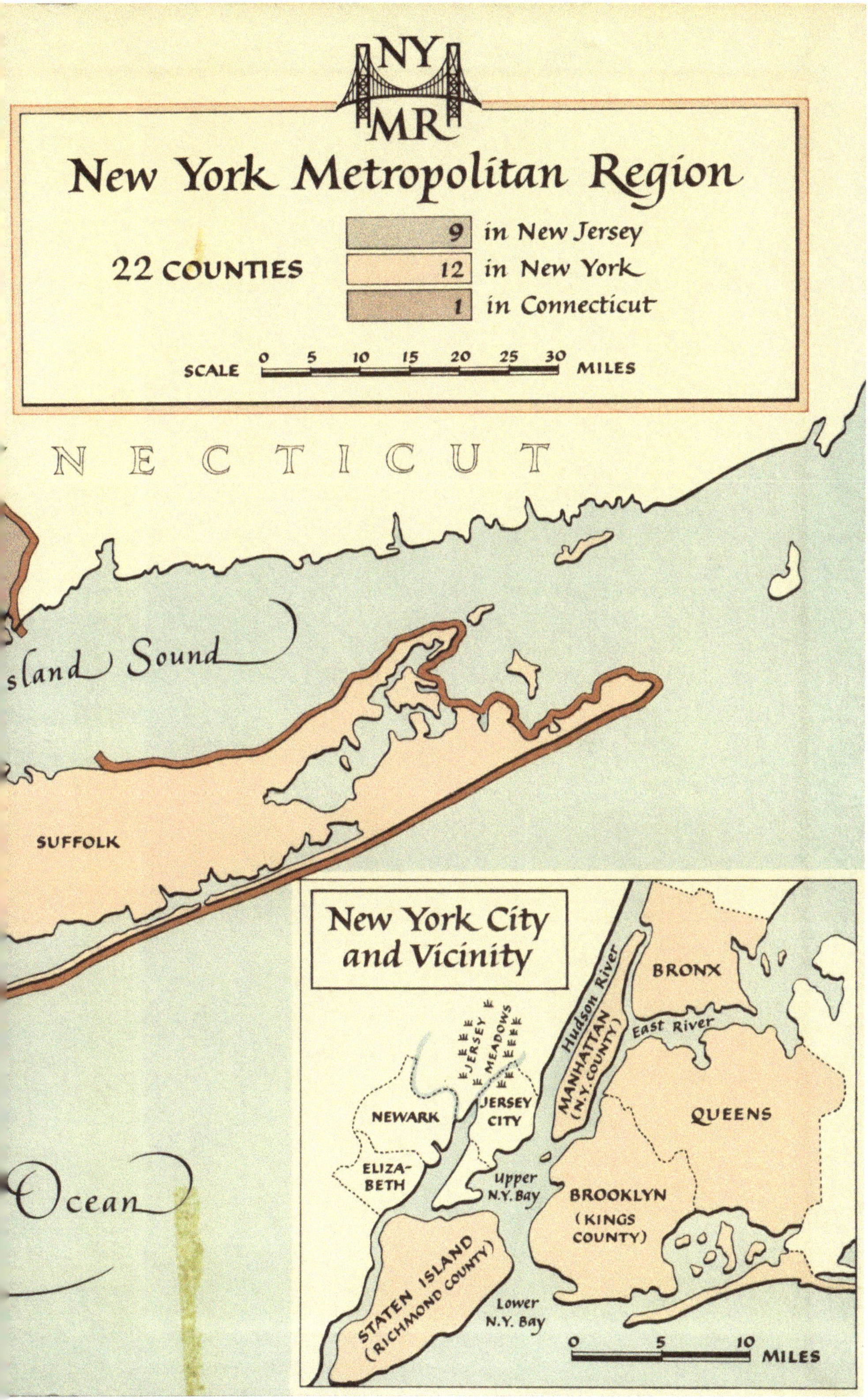
NY
MR
New York Metropolitan Region
22 COUNTIES
9 in New Jersey
12 in New York
1 in Connecticut
SCALE 0 5 10 15 20 25 30 MILES
NECTICUT
sland Sound
SUFFOLK
Ocean
New York City and Vicinity
Hudson River
BRONX
East River
MANHATTAN (N.Y. COUNTY)
JERSEY MEADOWS
JERSEY CITY
NEWARK
QUEENS
ELIZA-BETH
Upper N.Y. Bay
BROOKLYN (KINGS COUNTY)
STATEN ISLAND (RICHMOND COUNTY)
Lower N.Y. Bay
0 5 10 MILES

Bei Fragen zur Produktsicherheit wenden Sie sich bitte an:
If you have any questions regarding product safety,
please contact:

Walter de Gruyter GmbH
Genthiner Straße 13
10785 Berlin
productsafety@degruyterbrill.com